Maize in the Third World

WINROCK DEVELOPMENT-ORIENTED
LITERATURE SERIES
Steven A. Breth, series editor

Published in cooperation
with Winrock International Institute
for Agricultural Development

Maize in the Third World

Christopher R. Dowswell,
R. L. Paliwal,
and Ronald P. Cantrell

Routledge
Taylor & Francis Group

LONDON AND NEW YORK

First published 1996 by Westview Press

Published 2018 by Routledge
52 Vanderbilt Avenue, New York, NY 10017
2 Park Square, Milton Park, Abingdon, Oxon OX14 4RN

Routledge is an imprint of the Taylor & Francis Group, an informa business

A CIP catalog record for this book is available from the Library of Congress

ISBN 13: 978-0-367-01231-1 (hbk)
ISBN 13: 978-0-367-16218-4 (pbk)

Contents

Tables and Figures

Tables

Figures

Acknowledgments

During the long process of writing this book, many people helped us by generously giving their time and sharing their knowledge of maize science and maize production. In particular, we would like to thank Bob Havener, former director-general of CIMMYT and former president of Winrock International, for commisioning this book and for maintaining his faith that eventually it would see the light of day.

Scientists at CIMMYT, IITA, and in several national research institutes provided information and reviewed various drafts. In particular, we would like to thank David Beck, Kelly Cassaday, Hugo Cordova, Carlos de Leon, Greg Edmeades, Gonzalo Granados, Delbert Hess, David Jewell, John Mihm, Bobby Renfro, Sutat Sriwatanapongse, Suketoshi Taba, C. Y. Tang, Surinder Vasal, and Richard "Charlie" Wedderburn of CIMMYT and Nilsa Bosques-Perez, Les Everett, S. K. Kim, and Jim Mareck of IITA for their review and comment on different chapters. We would also like to thank Luis Manlio Castillo and Alejandro Ortega of ICTA in Guatemala and Sujin Jinahyon of Kasetsart University in Thailand for supplying information about their national maize programs. Recognition must also be given to the late Haldore Hanson for his perspectives on Chinese agriculture and to the late Johnson Douglas for his insights into seed production.

We owe a special debt of gratitude to Donald L. Duvick, who was the principal external reviewer of the manuscript. Not only did he offer many valuable suggestions for improvement, he gave us encouragement to finish the book.

Finally, we are especially grateful to Steven Breth of Winrock International who did so much to improve the organization, clarity, and accuracy of the text as well as to oversee the entire publication design and production process.

We express our thanks to the United Nations Development Programme and Winrock International for their support of the development of this book.

Despite the long and intensive review process that the book has gone through, the authors assume full responsibility for any errors and deficiencies that remain.

Christopher R. Dowswell
R. L. Paliwal
Ronald P. Cantrell

1

Maize in the World Economy

Importance of Maize

Among the world's cereal crops, maize ranks second to wheat in production, with milled rice third. However, among the developing economies, maize ranks first in Latin America and Africa but third after rice and wheat in Asia. Worldwide, maize is the most widely grown cereal crop. Seventy countries, including 53 developing countries, plant maize on more than 100,000 hectares. The far-reaching distribution of maize production is an indication of its excellent capacity to adapt to many environments (Fig. 1.1). Maize is grown at latitudes varying from the equator to slightly above 50° north and south, from sea level to over 3,000 meters elevation, under heavy rainfall and in semi-arid conditions, in cool and very hot climates, and with growing cycles ranging from 3 to 13 months.

Five hundred million tons of maize are produced annually on 130 million hectares (Table 1.1). Sixty-four percent of the world's maize area is found in developing countries, even though only 43 percent of world production is harvested there. The difference between the industrialized and developing countries is striking. The average yield for industrialized countries is 6.2 t/ha, compared with only 2.5 t/ha for developing countries.

The disparity in average yield levels is a consequence of environmental, technological, and organizational factors. Most industrialized maize-producing countries have benign temperate environments and employ input-intensive and highly mechanized maize production systems. In contrast, the developing countries generally have hotter, more difficult maize production environments and employ lower-input, and therefore, lower yielding maize technologies.

The 20 largest maize-producing countries account for 91 percent of world production and 80 percent of the total maize area (Table 1.2). The

FIGURE 1.1 Primary maize-growing areas (one dot = 75,000 tons).

United States produces 42 percent of world maize crop, followed by China with 19 percent.

Utilization of Maize

Maize has been put to a wider range of uses than any other cereal as a human food, as a feedgrain, a fodder crop, and for hundreds of industrial purposes because of its broad global distribution, its low price relative to other cereals, its diverse grain types, and its wide range of biological and industrial properties. The highest per capita maize utilization rates (Fig. 1.2) occur in countries where most of the grain is fed to animals or where maize is the preferred food staple.

About 66 percent of the global maize harvest is fed to livestock, 20 percent is consumed directly by humans, 8 percent is used in industrially processed food and nonfood products, and 6 percent is seed or wasted (Table 1.3). The sophistication of the maize economy is reflected in the industrial utilization data. In the developed economies of North America and Western Europe, industrial uses (food and nonfood) account for 16 percent of consumption, but in the developing countries

only about 2 percent of consumption is destined for industrial processing.

Feed Uses

Animal feed accounts for 70 percent or more of total maize utilization in industrialized economies, including Eastern Europe and the former Soviet Union, and in certain middle-income and newly industrialized nations of the Third World. In the United States in 1991, 150 million tons of maize—31 percent of total world production—were fed to livestock, either directly (20 million tons) or in formula feed in which maize is a major component (130 million tons). Among the developing countries, feed uses account for 77 percent of maize utilization in the Southern Cone of South America, 60 percent in North Africa and the Middle East, and 80 percent in the newly industrialized nations of the Pacific Rim. In contrast, low-income developing countries generally use less than 20 percent of their maize as feedgrain.

In the industrial economies, maize is the feed ingredient of choice in formula feeds because of its low cost and high degree of consistency. Despite the crop's biological superiority as an animal feed, maize is less common in formula feed in developing countries because their feed industries are less well developed and because in some regions, especially Asia, crops such as sorghum, millet, wheat, and rice are available at prices low enough to make them cost-effective substitutes for maize.

In the United States, about 10 percent of the maize area is harvested for silage, yielding about 100 million tons of biomass, while in northern Europe, about 30 percent of the area is harvested for silage. In Germany, for example, three fourths of the maize is used for silage. In some low-income developing countries, maize is an important green fodder for farm livestock, although statistics on this end use are not recorded. In Pakistan, India, and Egypt, for example, maize is more important as a fodder for water buffalo and other cattle than as a cereal grain for human consumption and feed uses.

Food Uses

More than half of all maize is utilized directly as a human food in the Andean countries of South America; Mexico, Central America, and the

TABLE 1.1 Average annual maize area, yield, and production, 1990-92.

Region	Area (million ha)	Yield (t/ha)	Production (million t)
Latin America	26.6	2.1	56
Mexico, Central America, & Caribbean	9.2	1.9	18
Andean region[a]	2.3	1.6	4
Southern Cone, South America[b]	15.1	2.3	35
Africa	20.6	1.4	30
East & Southern Africa[c]	12.9	1.4	18
West & Central Africa	6.5	1.2	6
North Africa	1.3	4.3	5
Asia	39.1	3.2	125
West Asia	1.0	3.2	3
South Asia	7.8	1.5	12
Southeast Asia	8.8	1.9	16
East Asia	22.1	4.6	102
Europe	13.8	4.6	63
Western Europe	3.9	7.2	28
Eastern Europe	9.8	3.5	34
USA & Canada	29.0	7.5	217
Former USSR	2.9	3.1	9
All developing	82.6	2.4	202
All industrialized	46.2	6.2	283
World	129.8	3.8	499

[a] Colombia, Ecuador, Peru, Bolivia.
[b] Argentina, Brazil, Chile, Paraguay, Uruguay.
[c] Includes South Africa.
Source: FAO Agrostat/PC files (1993)

Caribbean; Africa; and South and Southeast Asia. Maize accounted for at least 15 percent of the total daily calories in the diets of people in 23 developing countries, nearly all in Africa and Latin America (Fig. 1.3).

Industrial and Seed Uses

Worldwide, 14 percent of the maize crop is processed industrially for food, feed, and nonfood uses (Table 1.3). The most elaborate and diversified uses of maize occur in the United States where, in 1991, about 34 million tons of maize—19 percent of national production—were used for industrial purposes and seed processing (National Corn Growers Association 1992). Wet-milling of maize for starch and sweetener manufacture consumed 20 million tons. Three-fourths of this amount was for sweeteners, primarily for soft drinks. Maize was also an important raw material for ethanol fuel production, which consumed 10 million tons. Another 3 million tons were used in dry-milled and alkaline-cooked cereal products and baked products. Distilled products used 0.3 million tons, and hybrid seed sales accounted for 0.7 million tons.

TABLE 1.2 The 20 largest maize producers, 1990-92.

Country	Area (million ha)	Yield (t/ha)	Production (million t)
United States	28.05	7.5	210.7
China, Peoples Rep.	21.40	4.5	97.2
Brazil	12.64	2.0	25.2
Mexico	7.21	2.0	14.6
France	1.73	7.1	12.2
India	5.98	1.5	9.2
Former USSR	2.94	3.1	9.0
Yugoslavia	2.20	3.8	8.4
Romania	2.80	2.9	8.0
Argentina	1.97	4.0	7.8
Indonesia	3.23	2.2	7.0
South Africa	3.32	2.0	6.7
Italy	0.82	7.8	6.4
Canada	0.97	6.5	6.3
Hungary	1.10	5.7	6.3
Egypt	0.86	5.9	5.0
Philippines	3.60	1.3	4.7
North Korea	0.71	6.3	4.4
Thailand	1.45	2.6	3.7
Spain	0.46	6.5	3.0
Total	103.44	4.4	456.2

Source: FAO Agrostat/PC files (1993)

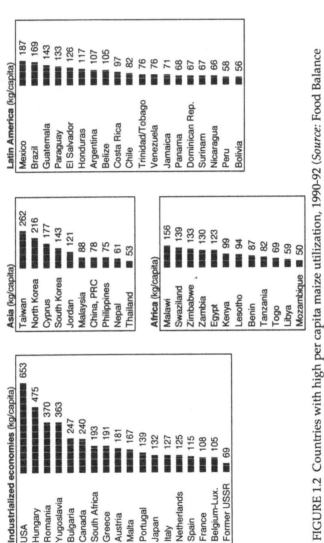

Industrialized economies (kg/capita)

Country	kg/capita
USA	653
Hungary	475
Romania	370
Yugoslavia	363
Bulgaria	247
Canada	240
South Africa	193
Greece	191
Austria	181
Malta	167
Portugal	139
Japan	132
Italy	127
Netherlands	125
Spain	115
France	108
Belgium-Lux.	105
Former USSR	69

Asia (kg/capita)

Country	kg/capita
Taiwan	262
North Korea	216
Cyprus	177
South Korea	143
Jordan	121
Malaysia	88
China, PRC	78
Philippines	75
Nepal	61
Thailand	53

Africa (kg/capita)

Country	kg/capita
Malawi	156
Swaziland	139
Zimbabwe	133
Zambia	130
Egypt	123
Kenya	99
Lesotho	94
Benin	87
Tanzania	82
Togo	69
Libya	59
Mozambique	50

Latin America (kg/capita)

Mexico	187
Brazil	169
Guatemala	143
Paraguay	133
El Salvador	126
Honduras	117
Argentina	107
Belize	105
Costa Rica	97
Chile	82
Trinidad/Tobago	76
Venezuela	76
Jamaica	71
Panama	68
Dominican Rep.	67
Surinam	67
Nicaragua	66
Peru	58
Bolivia	56

FIGURE 1.2 Countries with high per capita maize utilization, 1990-92 (*Source:* Food Balance Sheets, FAO Agrostat/PC files, 1993).

Global Trends

Production

From 1950 to 1980, world maize production increased (Fig. 1.4) from about 145 million tons to 450 million tons, growing at a faster rate than either wheat or rice. The 1980s, however, were a volatile decade for maize production. While production continued to grow steadily in the developing countries, the industrialized countries suffered sharp reductions in 1983 due to severe drought. More normal weather patterns in 1984 and 1985 helped production recover, but poor weather returned in 1986, 1987, and 1988 and harvest were again small. By the end of the decade, the growth of world production resumed and output exceeding 500 million tons in 1992.

But in 1991-92 the worst drought of the 20th century afflicted South Africa and Zimbabwe, which had usually escaped severe droughts in the past. Southern Africa suffered 75 to 80 percent declines in maize production in 1992 and East Africa had 10 to 20 percent declines. The political and economic upheavals in the former Soviet Union and Eastern Europe nations also severely depressed maize production. In contrast, in 1992 the United States had an all-time record maize harvest of 229 million tons.

TABLE 1.3 Estimated maize utilization in different regions, 1989-91.

Region	Direct use (%)		Industrial use (%)		
	Food	Feed	Food	Nonfood	Other[a]
USA & Canada	2	79	8	8	3
Western Europe	10	80	2	4	4
E. Europe & USSR	8	72	2	6	12
Mex., Cent. America, Caribbean	72	15	4	2	7
Andean region, South America	60	30	2	2	6
Southern Cone, South America.	13	77	2	2	6
East & Southern Africa	71	19	1	1	8
West & Central Africa	74	9	6	2	9
North Africa & Middle East	30	60	1	3	9
South Asia	83	3	1	4	9
Southeast Asia	57	36	1	3	3
East Asia	25	67	1	3	4
Industrialized economies	5	79	5	8	3
Developing countries	40	50	1	1	8
World	20	66	4	4	6

[a] Seed use and wastage.

Source: FAO Agrostat PC files (1992)

Production Growth

Over the past four decades, the relative contributions of yield increase and area expansion to growth in maize production have differed considerably between developing countries and the industrialized countries (Fig. 1.5). In the industrialized countries, more than 90 percent of the growth in maize production can be attributed to the adoption of yield-increasing technologies. In the developing countries, area expansion has accounted for about half of the growth in maize production, but yield-increasing technologies are becoming more important.

The rates and sources of higher world maize production in past two decades vary considerably from region to region (Table 1.4). The highest growth rates in maize production occurred in the developing world, where average annual production grew by 3.8 percent during the 1970s and 3.0 percent during the 1980s. The weakest production gains were recorded in Eastern Europe and the·former Soviet Union, where the maize production growth of the 1970s (2%/year) was wiped out during the 1980s (-2.5%/year). In the developed market economies, maize production grew by 3.6 percent per year during the 1970s. This strong production growth was not sustained during the 1980s, as governments in

Country	Calories from maize relative to total diet (%)
Zambia	56
Malawi	54
Lesotho	46
Guatemala	45
Mexico	44
Kenya	44
South Africa	40
Zimbabwe	37
El Salvador	36
Swaziland	36
Kenya	36
Honduras	34
Cape Verde	30
Namibia	30
Tanzania	26
Nicaragua	26
Botswana	25
Egypt	25
Benin	20
Nepal	17
Somalia	17
Reunion	17
Ethiopia	15

FIGURE 1.3 Developing countries where maize accounts for over 15 percent of total calories in diet, 1989-90 (*Source:* Food Balance Sheets, FAO Agrostat/PC files, 1993).

the industrialized countries lowered price subsidies in an attempt to reduce crop surpluses.

In the developing world, area expansions have been strongest in sub-
Saharan Africa and Southeast Asia. An analysis of these statistics re-
veals that yield increases are becoming ever more the primary source of
production growth.

Changes in Maize Area. Between 1969-71 and 1989-91, the global
maize area grew from 104 to 129 million hectares, a 24 percent increase.
In the developed market economies, maize area increased by 2 million
hectares, with most of the growth in Europe and Canada, while in Eastern Europe and the former Soviet Union, it decreased by 4 million hectares. However, in the developing countries, the maize area increased
by 27 million hectares over these three decades: 5.5 million hectares in
Latin America, 10 million in sub-Saharan Africa, 5 million hectares in
South and Southeast Asia, and 6.5 million hectares in East Asia. The
maize area remained relatively unchanged in North Africa and the
Middle East. During the 1970s and 1980s, the rate of expansion in maize
area has declined in most regions and countries. The annual growth rate
in maize area during the 1980s was one-fifth the rate of the 1960s.

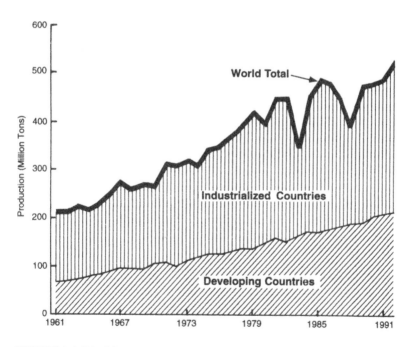

FIGURE 1.4 World maize production 1961-93. (*Source:* FAO Agrostat
PC files).

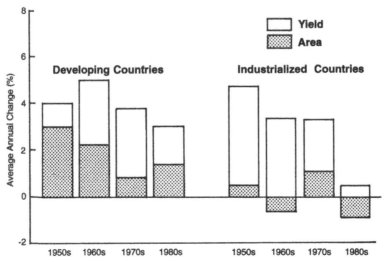

FIGURE 1.5 Contribution of changes in area and yield to growth of maize production in developing and industrialized countries, 1950-90 (*Source:* CIMMYT 1992).

In low-income countries where maize is a dietary staple, the driving force for area expansion has been rising maize demand due primarily to population growth. This trend is notable in much of sub-Saharan Africa and in parts of Latin America where yields have been stagnant and where considerable land resources are still available to bring into maize production. In contrast, area expansion in nontraditional maize-producing environments is generally the result of advances in maize research and better access to improved maize technology. For example, maize researchers have been able to develop early maturing, cold-tolerant maize genotypes, which have allowed maize to be profitably grown at higher latitudes in Europe and China. Similarly, the development of maize genotypes tolerant to soils that are acidic and have high levels of free aluminum has allowed maize to be grown in areas such as the Cerrado, the savanna lands of Brazil. And overcoming serious disease problems, such as downy mildew in Southeast Asia, has opened large new areas to maize production. Over time, however, as land has become more limiting, expansion of maize area is becoming less important than improvements in maize yields as the source of increased production.

Changes in Maize Yields. In the more-industrialized countries, increasing yields were the primary source of higher maize production during the 1970s, although the growth in yields has tended to slow with each passing decade. Growth in yields was strong in Eastern Europe

TABLE 1.4 Growth of maize area, yield, and production, 1971-80 and 1981-90.

| | Growth rate (%/yr) | | | | | |
| | Area | | Yield | | Production | |
Region/Groupings	1971-80	1981-90	1971-80	1981-90	1971-80	1981-90
East & Southern Africa	0.4	2.8	0.4	1.0	0.8	3.8
West & Central Africa	-1.0	5.5^a	0.0	1.7	-1.0	7.2*
North Africa	0.9	-0.1	2.0	3.8	2.8	3.7
West Asia	0.2	-1.4	2.5	5.3	2.7	3.9
South Asia	0.3	0.5	1.0	2.3	1.3	2.8
Southeast Asia	2.5	2.0	2.4	2.6	4.9	4.6
East Asia	2.7	1.4	4.3	2.7	7.0	4.1
Mexico, Central America, Caribbean	-1.3	-0.1	3.4	0.2	2.1	0.1
Andean region, South America	-1.4	4.4	1.9	0.6	0.5	5.0
Southern Cone, South America	0.0	-0.5	1.4	0.2	1.5	-0.2
Developing countries	0.8	1.3	3.0	1.7	3.8	3.0
Eastern Europe & former USSR	-1.2	-0.1	3.2	-2.5	2.0	-2.5
Developed market economies	1.7	-1.1	1.9	1.2	3.6	0.1
World	0.9	0.5	2.6	0.4	3.5	0.9

^{a} FAO's West Africa statistics are highly questionable due to an error that led to an underestimation of the Nigeria maize area by 1 million hectares during the early 1980s. Correction of this error has resulted in overestimations of actual rates of growth in area in this region.

Source: FAO Agrostat/PC files (1991)

and the former Soviet Union during the 1970s (3.0%/year), a period of rapid adoption of hybrids and fertilizers. In sharp contrast, maize yields shrank by 2.5 percent per year during the 1980s, as these economies became unsettled. Yield increases in the developed market economies were especially strong during the 1950s and 1960s. Even so, maize yields continued to rise by 1.9 percent per year during the 1970s.

Although variable from region to region, the rates of increase in average maize yields in developing countries show a strong upward trend in recent decades as Third World maize farmers have gained access to fertilizers and high-yielding varieties and hybrids. Strong growth in average maize yields in the developing countries (3.0%/year) during the 1970s slowed during the 1980s, largely because of the general economic crisis that gripped many developing countries.

Maize Use

Growth in global maize utilization has been driven by the rapidly increasing demand for maize as a livestock feed in developing countries and for industrial food and nonfood products. In general, direct food uses of maize have declined as per capita incomes and milk, meat, and egg consumption have increased. The demand for maize as a livestock feed during the past two decades has expanded the most in the middle-income countries (primarily oil-exporting) and in the newly industrialized nations (primarily Pacific Rim countries, Mexico, and Brazil), averaging 4 percent per year.

In regions where incomes have grown rapidly, food consumption patterns have changed, leading to sharply increased demand for livestock and poultry products and consequently for feedgrains. Considerable economic activity and investment have been associated with the development of intensive livestock, poultry, and feedgrain industries in most of Latin America, North Africa, and the Middle East, and in Pacific Rim countries.

In the United States, the industrial food and nonfood (including seed) uses of maize have increased from 14 million tons in 1975 to 34 million tons in 1991 (National Corn Growers Association 1992), with the maize destined for sweetener production increasing from 5.3 to 15.6 million tons and the maize used in ethanol production increasing from 0.6 million tons to 9.9 million tons.

Maize Trade

Growth in maize production, especially in the more favored temperate environments where hybrids and high-yielding agronomic practices are used, has greatly expanded the availability of maize in world markets. In 1950 world maize trade stood at about 16 million tons. By 1980, it had increased to a peak of about 80 million tons, after which rising maize production in developing countries and shortages of foreign exchange in many countries tended to dampen maize trade. In the 1990s, the volume of maize traded has fluctuated between 60 and 70 million tons. Virtually all the maize moving in world markets is destined to be fed to livestock and poultry.

During 1990-92, the major exporters were the United States (67% of the total), China, France, Argentina, Hungary, and Thailand (Table 1.5). These six countries accounted for 95 percent of world maize exports.

China is a new exporter. It prefers to sell maize to its Asian neighbors to help offset the cost of importing wheat. In 1993 China exported 11.5 million tons of maize. Thailand, which had been a major exporter since the mid-1970s, decreased it maize exports considerably in the 1990s. One reason is that its primary client, Japan, wanted better control of aflatoxin on the grain. Also in recent years, Thailand has experienced a boom in production of poultry, which is raised on Thai maize. Thus, rather than exporting maize, Thailand today exports increasing quantities of poultry meat and eggs to Pacific Rim countries.

Among the maize-importing countries, 22 imported more than 500,000 tons annually during 1990-92. Together they accounted for 75 percent of world maize imports (Fig. 1.6).

Overall, industrial economies accounted for 80 percent of global maize imports during 1990-92. Among developing countries, over 10 imported more than 100,000 tons of maize per year. Most were middle-income and newly industrialized countries.

Maize Prices

Between 1961 and 1993, world maize supplies expanded more rapidly than demand, putting downward pressure on the price of maize (adjusted for inflation). Gains in productivity, owing to the use of improved varieties and agronomic practices, have played a major role in lowering maize prices. Because of the predominance of the United

TABLE 1.5 Countries exporting over 500,000 tons of maize annually, 1990-92.

Country	Exports (000 t)
USA	46,655
China, People's Rep.	7,167
France	6,338
Argentina	4,328
Hungary	914
Thailand	871

Source: FAO Agrostat PC files (1993).

Country	Imports (000 t)
Japan	▮▮▮▮▮▮▮▮▮▮▮▮▮▮▮▮▮▮▮▮▮▮▮▮▮▮▮▮▮▮▮ 16,345
Former USSR	▮▮▮▮▮▮▮▮▮▮▮▮▮▮▮▮▮▮▮▮▮ 11,376
South Korea	▮▮▮▮▮▮▮▮▮▮▮ 6,082
Taiwan	▮▮▮▮▮▮▮▮▮▮ 5,422
Mexico	▮▮▮▮▮ 2,277
Netherlands	▮▮▮▮ 1,935
Spain	▮▮▮▮ 1,759
United Kingdom	▮▮▮▮ 1,612
Malaysia	▮▮▮▮ 1,587
Egypt	▮▮▮ 1,548
Belgium-Lux.	▮▮ 1,094
Italy	▮▮ 1,002
Iran	▮▮ 969
Algeria	▮▮ 906
Portugal	▮▮ 784
Brazil	▮▮ 653
Germany	▮ 606
Venezuela	▮ 580
Saudi Arabia	▮ 571
Peru	▮ 545
Romania	▮ 517
Mozambique	▮ 501

FIGURE 1.6 Countries importing an average of over 500,000 tons of maize annually, 1990-92 (*Source:* FAO Agrostat/PC files, 1993).

States in the maize trade, the best measure of world maize prices is the f.o.b. U.S. price (Fig. 1.7). Since 1960, real U.S. maize prices, adjusted for inflation, have declined steadily, except during the world food shortages of 1973-76, resulting from the OPEC oil embargoes and several years of bad weather. Severe production shortfalls in southern Africa and East Africa in 1991-92 are likely to be translated in increased demand for imports, which could exert upward pressure on global prices. On the other hand, the United States harvested a record maize crop in 1992, which will probably compensate for shortages in Africa.

Maize price trends over the past 30 years have paralleled those of wheat, but the price of maize has fallen relative to the price of rice. In the mid-1980s, maize prices, adjusted for inflation, reached record low levels in world grain markets (Fig. 1.7). Since 1988, however, nominal maize prices have rebounded, although the trend has continued downward.

The long-term trends in maize prices indicate that both supply and demand for maize have changed at about the same rate. On the supply side, cost-reducing technologies, particularly in the developed countries, has led to the doubling of maize production without increases in real prices. At the same time, increased demand for maize, especially as an animal feed, generated by rising incomes and population growth, has absorbed the increased production.

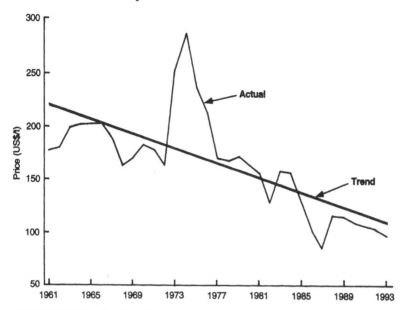

FIGURE 1.7 Real price of maize in world markets (1990 U.S. dollars).

Market interventions by governments in important maize-producing and maize-consuming countries had important—and often adverse—influences on international maize prices. In developed countries, protectionism generally favors agriculture and is paid for by taxpayers, consumers, and producers of other tradable products in the economy. In the developing countries and in the centrally planned economies, in contrast, protectionism penalizes agriculture, in favor of urban consumers and other sectors of the economy.

In response to pressure from the World Bank, IMF, and various bilateral donor agencies, governments in most developing countries are removing discriminatory maize price policies. In the United States and the European Union, direct and indirect price supports have had a significant depressing effect on the international maize price. These declining maize prices have worsened the economic situation of agriculture in maize-exporting countries, notably Argentina and Thailand, which has had an adverse effect on farmers.

Mexico. Photo courtesy of CIMMYT.

2

The Maize Plant and Its Uses

Origin of Maize

Maize* (*Zea mays* L.) is a member of the grass family, Gramineae, to which all the major cereals belong. Cultivated maize is a fully domesticated plant, that is, unable to survive without human husbandry. It has the highest grain yield potential of all the cereals and is a wonder of efficiency in transforming the energy of the sun into food energy.

The origin of maize is controversial (Galiant 1988). Annual teosinte (*Zea mexicana*) is an ancestor and the closest relative of maize. Like *Zea mays*, it has 10 pairs of chromosomes, which are cytogenetically similar to those of *Zea mays*, and the plants hybridize freely, producing fertile progeny. Both maize and teosinte have the male flowers borne in the tassel and the female flowers borne on a lateral branch. The species differ chiefly in the nature of their female organs. Maize depends upon humans to separate and disperse the seed from the highly protected ear. Teosinte is a self-sowing plant whose seed shatters at maturity and is dispersed.

How long maize has been grown and where it originated are still subjects of considerable speculation. The oldest evidence of domesticated maize comes from archaeological sites in Mexico where small cobs estimated to be 7,000 years old have been found in caves. This estimate compares well with the generally accepted dates for the origin of agriculture in both the old and new worlds, 8,000 to 10,000 years ago.

From its center of origin in Mexico and Central America, maize migrated to the rest of Latin America, the Caribbean, and the United States and Canada. One theory holds that maize was also domesticated independently in the Andean countries of South America by ancestors of the

---

* Most of the English-speaking world refers to *Zea mays* as maize, except for the United States, the largest producer, where the common name is corn. This usage traces to the historic British practice of denoting any area's principal cereal crop as "corn."

Incas. European explorers of the Americas took maize to Europe, and traders later took it to Asia and Africa.

Evolution and Spread of Maize

The evolution of maize has been strongly affected by the hand of man, or more probably, woman. Between 5000 BC and 1000 AD, as a result of favorable natural mutations and hybridization among different races of maize and specific ear selection by farmers for size, color, and kernel characteristics, maize evolved from a plant having a small, self-sowing wild type of ear into a plant having a multi-rowed, highly productive, husk-enclosed ear that requires the help of humans to disperse its seeds for propagation and survival.

By 1000 AD, maize types had developed that looked like modern maize varieties. Most of the improvement undoubtedly was done by simple mass selection—cultivators retained the largest and most desirable ears to use as seed in the next season. Variation among ears due to environmental effects obscured genetic differences in yield, plant height, and other quantitative traits so that the rate of improvement from this selection process was very slow. Also, the source of pollen was not controlled so that contamination from other maize varieties frequently occurred. In some cases, this hybridization proved to be very fortunate, leading to the creation of vigorous new varieties. By the time Columbus encountered maize in Cuba in 1492, natives from Chile to Canada were growing it.

The explorers and traders of the 16th century were primarily responsible for the introduction of maize beyond the Americas. Probably maize first strayed beyond the Americas through Columbus, who carried kernels of several Caribbean yellow flint landraces from Cuba to Spain when he returned from his first voyage in 1493. It was maize's high multiplication ratio and high yield potential that led to its spread through Europe, Asia, and Africa. In Europe, within a hundred years, maize had gone from being a botanical curiosity, to a kitchen-garden vegetable, to a major commercial field crop. Though it took more than 200 years to take hold throughout Europe, maize was grown widely in Spain, Italy, and southern France by the end of the 1500s.

Maize appears to have reached Africa in the early 1500s by way of Portuguese traders. Maize started displacing other food crops, and by the 1700s rose to prominence among food plants in West and Central Africa, especially in the Congo Republic, Benin, and western Nigeria.

Maize also arrived in South Asia in the early 1500s, probably by both land and sea. Portuguese traders, and probably Arab traders from Zanzibar, introduced maize along the western coast of India. It was also introduced into what is now Northwest Pakistan by traders working along the silk route. Maize took hold in the highland areas of the Himalayas, where the multiple uses of the plant and grain fit well with the farming needs of people in these areas.

Maize had reached Southeast Asia by the mid-1500s and by the mid-1600s it had become a well-established crop in Indonesia, the Philippines, and Thailand. According to some accounts, maize reached the Philippines from the west before Magellan reached these islands from the east in 1521.

By the mid-1700s, maize also was widely grown in the southern Chinese provinces of Hunan, Szechwan, and Fukien. Rapid expansion in the population during the 18th century made it necessary to clear more land in the hills and mountains outside the rice-growing plains. Thus, necessity forced the Chinese to turn to new foods to meet their dietary needs, and maize became widely established in the north of China and spread to Korea and Japan.

American farmer-seedsmen made significant improvements in maize in the 19th century and developed some outstanding open-pollinated varieties. These farmer-seedsmen began to study maize closely and intentionally cross-pollinated different races to develop improved varieties. The inter-racial hybridization of hard flint varieties from New England and soft, large-kernel, dent varieties from the south resulted in the initial high-yielding Corn Belt dent varieties. By the mid-1800s, a number of outstanding improved open-pollinated varieties had been developed through this recombination of races.

In the 20th century, intensive research in plant breeding led to spectacular improvements in yield potential. Advances in breeding methods have been accompanied by increased knowledge of plant morphology, of pollination, and of the genetics and cytology of maize. Maize is probably the most thoroughly studied of all the plant species. The development of high-yielding hybrids revolutionized maize production in the U.S. Corn Belt and later in other countries. Hybrid maize remains the greatest practical achievement of plant genetics to date.

Maize Plant Growth

Modern maize has no equal among other cultivated species, or for that matter, in the wild as an efficient producer of grain. The kernels on an ear of maize cling tightly to the rigid cob, the creation of centuries of selective breeding. If the ear were to drop to the ground, so many competing seedlings would emerge that in all likelihood, few would grow to maturity. Plant growth and development can be divided into five major phases, each with unique characteristics (Shaw 1988):

Planting to emergence. Plant growth between germination and seedling emergence depends on soil moisture, temperature, soil aeration, and seed vigor. Before germination, the planted seed absorbs water and swells. With warmer temperatures, germination will start earlier and proceed faster, assuming sufficient soil moisture is available. With adequate moisture and proper planting depth, seedlings will emerge in about 8 days when daytime soil temperatures are optimum—about 20°C. Seedlings take more than 2 weeks to emerge at temperatures of 10°C. Below this temperature, germination is sharply lower.

Shortly after emergence, an important change takes place. The plant shifts from a dependence on starch stored in the seed to self sufficiency. During the early part of its life, the maize plant requires a limited amount of moisture. Stress immediately after emergence decreases the starch and chlorophyll content of the seedlings, but it tends to result in deeper penetration into the soil of the roots, which makes the plant better able to withstand dry weather later, thus usually offsetting any immediate detrimental effects of stress.

Emergence to tasseling. The vegetative period—when the photosynthetic factory is established and begins to operate at full capacity—lasts from emergence to tasseling. Until about six leaves have fully emerged, the plant's growth point remains below the surface, allowing the plant to withstand a moderate frost. At that temperature, although aboveground parts are killed, recovery is usually rapid and almost complete, with no appreciable effect on yield.

Later in the vegetative stage, the relationship between weather and yield becomes much more significant. In this stage, the plant grows very rapidly and water use is greater. Moisture stress during this period will cause yield reductions, and fertility stress will accentuate the yield losses. Flooding (water covering the roots for more than 24 hours) will also depress yields, especially if soil nitrogen levels are low.

Silking. The critical stage during which the silks are pollinated may last 1 to 2 weeks in the field. During this time, each silk must emerge for pollination if a kernel is to develop. The silks grow from 2.5 to 4 centimeters a day and continue to elongate until fertilized. Pollination occurs when the falling pollen grains are caught by the new moist silks. A captured pollen grain takes about 24 hours to grow down the silk to the ovule where fertilization occurs, after which the ovule develops into a kernel. Generally, 2 to 3 days are required (unless there is moisture stress) for all silks on a single ear to be exposed and pollinated. Both drought and fertility stress at this stage retard the development of ovules, and drought may dry out the silks. As a result many ovules will be unfertilized or abort early, and therefore do not produce kernels.

Silking to maturity. Between silking and maturity, the grain is produced, and starch rapidly and steadily accumulates in the watery endosperm of the kernel. During the grain-filling period, the plant's dry-matter accumulation takes place almost entirely in the kernels. Moisture stress during grain-filling can reduce the final size and weight of the kernels. Although the effects on yield are not as severe as those of moisture stress during tasseling and silking, they can be significant. As the kernels become more mature, the potential yield reduction from stress declines. Varieties differ in their maturity period, however mild temperatures can lengthen it and high temperatures can diminish it.

Dry-down period. After physiological maturity, the dry-down period begins. Black layer development in the kernel signals the end of grain-filling. Kernels often contain 30 percent or more water when physiologically mature. After black layer development, the grain should dry down to 15 percent moisture or less before it is placed in storage. The rate of drying of the grain is affected by weather and the variety characteristics. Rainfall early in the drying stage is a major contributor to high kernel moisture content at harvest.

Types of Grain

Although all types of maize belong to the same species, the size, texture, shape, and color of the kernels on an ear of modern maize vary widely from race to race. This tremendous diversity is the result of centuries of selection, mutations, and hybridization. Several types of grain are recognized (Fig. 2.1).

Flint Maize

In flint maize, the entire outer portion of the kernel is composed of "hard" starch, which does not easily form a paste with water. The starch composition gives the kernel a shiny surface. Flint maize makes a good quality cornmeal (dry milling). There is less risk of spoilage in shipping and storage than with dent maize because the hard kernel absorbs less moisture. Flint grain is also more resistant to fungi and insect damage. Compared with dent maize, most flint maize varieties mature earlier, and their seeds germinate better in cold, wet soils. Flint maize grown at higher latitudes adapts better than other forms of maize. Flints come in many colors—white, yellow, red-blue, or variable.

Dent Maize

Dent maize is the most widely grown type. In the United States, it accounts for 95 percent of all maize. In dent maize, the hard starch is

Endosperm Type	POP	FLINT	DENT	FLOUR	SWEET
Photograph (natural size)	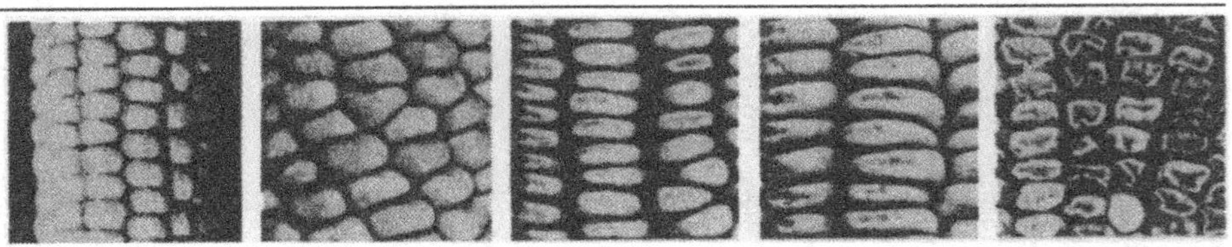				
Pericarp	very thick	thick	medium	stretched thin	thick-medium
Endosperm (mature)	hard	mostly hard	hard and soft	soft	glassy
Crown appearance (mature)	pointed or rounded	rounded	dented	slightly dented	wrinkled
Distribution	USA (Indiana) sporadic in all regions.	Argentina, Southern Europe, and marginal areas where storage and germination is difficult.	Worldwide	Latin America, American Southwest.	North America
Importance (New World)	<1%	14%	73%	12%	~1%
Use	confection	general	livestock feed industrial processing millground meal	*major:* direct human use from handground meal. *minor:* direct at milk stage,' parched and beverage.	*major:* direct at milk stage *minor:* parched and beverage.

Legend:
- hard (flinty)
- soft (granular)
- sugar (glassy)
- germ

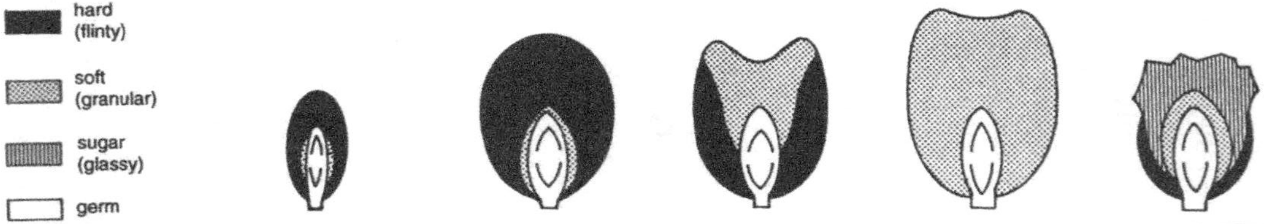

FIGURE 2.1 Principal kernel types of maize (Reproduced by permission of the publisher from *Maize*, Ciba-Geigy Ltd., Basle, Switzerland, 1979).

confined to the sides of the kernel. The amylose, or "soft" starch, which forms the core and cap, contracts when the grain is dried, producing the characteristic dent in the top of the kernel. Most dent maize is used for livestock feeding. Dent maize may be yellow, white, or red. Most of the world's commercially grown dent maize is yellow, although white is preferred as a food in sub-Saharan Africa, Mexico, Central America, and the Caribbean.

Sweet Corn

Sweet corn is grown primarily as a food and is harvested with about 70 percent moisture before hardening and drying of the grain starts. It is primarily grown in the United States and Canada. Yellow is the predominant grain color. The kernels are high in sugar when they are eaten on the cob, canned, or frozen. Sweet corn is a good source of energy. Twenty percent of the dry matter in sweet corn is sugar, compared with only 3 percent in dent maize at the green ear stage. Yellow grain sweet corn is also a good source of vitamins C and A. Unless it is eaten quickly after harvesting—or is processed—sweet corn loses its flavor. Canned and quick frozen sweet corn last almost indefinitely.

Floury Maize

Floury maize is grown in Andean highlands of South America and in the drier areas of the U.S. southwest, and various genotypic variations are used for selective food preferences. Floury maize is one of the oldest types of maize and was preferred by native populations because the kernels are easily ground to make flour. The kernels of floury maize are composed largely of soft starch and have little or no hard starch.

Popcorn

Popcorn is a popular snack food in many parts of the world. It is an extreme form of flint maize, although the kernels of popcorn are smaller. When heated to about 170°C, the grains swell and burst, turning inside out. At this temperature, water held in the starch in the kernel tissue turns to steam, and the pressure causes the endosperm to explode.

Waxy Maize

Waxy maize is named for the somewhat waxy appearance of the kernels. China was the original source of the waxy gene (*wx*), although waxy mutations have occurred in U.S. dent strains. Waxy maize starch is composed entirely of amylopectin, in contrast with common maize starch, which is approximately 78 percent amylopectin and 22 percent amylose. Waxy maize hybrids have been developed and are being grown to supply raw materials for specialty products of the wet-milling starch industry.

Biochemical Characteristics

Maize grain is palatable, nutritious, and rich in energy-producing carbohydrates (starch) and fats. It is higher in fat than rice or wheat (4% versus 2%). The energy value of maize (cornmeal, whole ground) is 3,578 calories per kilogram, compared with 3,629 calories for brown rice and 3,327 calories for whole wheat grain. Although maize contains about 10 percent protein, about half of the protein consists of zein, which is especially low in lysine and tryptophan, amino acids that are essential for single-stomach creatures, such as humans and pigs. Maize is also deficient in minerals, particularly calcium, and in the vitamin niacin. White maize is low in carotene, the precursor of vitamin A. Yellow maize has higher carotene content, but it decreases with time in storage.

In areas where people have diets that are heavily dependent on maize, weaning infants can develop the disease kwashiorkor, primarily due to protein deficiency, and blindness may also occur.

Storage Qualities

Maize grain has generally good storage properties. Moisture content is the principal influence on storability because the maize kernel is fairly hygroscopic. In most climates, freshly harvested grain has a moisture content of about 30 percent. In this state, the grain cannot be kept for any length of time. Instead, the grain must first be dried, preferably to 15% moisture content.

Safe kernel moisture levels for storage are influenced primarily by the relative humidity of the atmosphere within the storage area (environmental relative humidity, or ERH) and, to a lesser extent, by the temperature of the grain and the storage area. In tropical climates, maize that has a moisture content below 15 percent can be safely stored for prolonged periods in containers or structures that have an ERH of less than 60 percent and temperatures up to 30°C.

Maize stored with higher moisture content is perishable, owing the development of harmful yeasts, molds, and bacteria. Of the several fungi that commonly grow in stored grain, the most dangerous is *Aspergillus flavus*. This species produces aflatoxin, a mycotoxin that is hazardous to humans and animals. Molds that do not produce toxins can still be a problem, because they cause the grain to lose its palatability and nutritional properties. Further, spontaneous heating caused by kernel respiration accelerates the deterioration process. Wet grain containing 40 percent moisture begins to generate heat and decompose, releasing a sour odor, after only about 10 hours.

Rats, mice, and other rodents are an important source of damage to stored maize grain. They cause losses by consuming grain, by fouling grain with feces, urine, hairs, and dead bodies, by transmitting disease through body excreta, and by destroying containers and bags.

Insects can cause severe losses in stored maize grain. The degree of insect damage varies according to the grain type and adequacy of husk cover. Insects thrive on soft, starchy kernels and more readily infest maize ears in the field that are not completely covered by the husk, which is a frequent shortcoming of improved maize varieties. Some insects are able to penetrate the pericarp and lay eggs inside the kernel. The growing larvae thus feed on the inside of the kernel. Other insects follow these first attackers, feeding on kernels that are broken or have cracked pericarps, but usually not attacking healthy, undamaged kernels. A third group of insects feeds on broken grains, grain dust, and powder left by the other insect groups. The primary factors favoring insect infestation of a grain storage facility are availability of air (oxygen), high moisture content in air and grain, and warm temperatures.

Among the more common pests of stored maize are the rice weevil (*Sitophilus oryzae*) and the maize grain weevil (*S. zeamais*). These grain weevils infest the grain while it is still in the field, and then multiply in storage. The ear borer (*Mussidia nigrivenella*) also attacks the grain during the interval between physiological maturity and drying. Once grain moisture has been reduced to 13 percent, the level at which grain can be safely stored, it is no longer subject to ear borer attack.

Other insects, like the larger grain borer (*Prostephanus truncatus*), attack the crop after the grain has been stored, especially in facilities where the availability of food, air, moisture, and heat allow the insects to thrive. The larger grain borer, found in the Americas and Africa, is omnivorous—it attacks not only stored cereals but other crops such as cassava. In its native habitat, Central America, this pest even damages wood. In areas where the larger grain borer is present, it is important to dry ears rapidly, so that shelling can commence earlier. Once maize has been shelled, it is much safer from attack by this pest.

Uses of Maize

Maize is used in more ways than any other cereal. It is used as a human food and feed for livestock, for fermentation, and for industrial purposes. Every part of the plant has economic value. The grain, leaves, stalk, tassel, and even the cob, are used to produce hundreds of food and nonfood products (Watson 1988).

In Human Diets

In Mexico and Central America, the home of maize, the most common food product is tortillas. These are prepared by steeping maize kernels in a mixture of water and lime to remove the outer layer of the kernel and then grinding the dehulled kernels into dough. The lime solution improves flavor, imparts a dough-forming character, and increases niacin availability. The dough (*masa*, in Spanish) is then fashioned into round thin cakes and cooked on a heated clay or metal plate. Tortillas are eaten along with meats, sauces, and vegetables. Other food products consumed in this region include roasted or boiled green ears (ears harvested before physiological maturity) and various preparations made from ground grain.

In Caribbean countries, maize is commonly fermented before being used as the main ingredient in soups and broths. In Haiti, products from yellow maize are made from cracked grain or a very rough maize flour.

In the Andean countries of South America, maize is prepared in a variety of ways. Roasting green maize ears, popping the kernels, and toasting of large floury kernels are common. The Peruvian Indians developed the floury maize type for preparing *kancha*, a favorite food. They also developed the predecessors of modern sweet corn, and they brew a native maize beer, known as *chicha*.

In East Africa and southern Africa, maize is the staple food, and it is consumed as a thick porridge. The maize kernels are pounded and cooked with water and eaten, either in paste or cake form, with other foods. *Nshima* is a food made from maize meal mixed with relish and sometimes meat, fish, or legumes. In West Africa, *kenkey* is prepared from kernels that are soaked, dehulled, ground while wet, and fermented in water for 1 to 3 days before being steamed and served, usually with spices.

In Asia, maize forms a much smaller component of diets. It is usually cooked like rice—the cracked maize kernels are boiled and eaten with various dishes. It is also consumed steamed or roasted as green cob maize.

The most varied uses of maize as a human food occurs in the United States. Over 1,000 different items on the shelves of a typical supermarket are derived partially from maize. Such items include tortillas, maize flours, thickeners and paste (from maize starch), syrups, sweeteners (high fructose and dextrose), grits, breakfast cereals, chips, cooking oils, beer, and whiskey.

Feed Uses

Sixty-six percent of all maize—nearly 330 million tons a year—was fed to poultry and livestock during 1990-92 making maize the world's leading feedgrain. Animals eat it readily. Of all the cereal grains, maize gives the highest conversion of dry substance to meat, milk, and eggs. Because of its high starch and low fiber content, maize is one of the most concentrated energy food sources. Its low price and high energy value—both payoffs from research—have made maize the feed ingredient of choice in formulated feeds.

In the United States, maize growers retain nearly half of the harvest for use on their farms as a fattening ration for pigs and beef cattle. Maize is fed as a whole shelled grain, cracked shelled grain, or steam-flaked grain that is supplemented with various protein sources, minerals, and vitamins, using least-cost mixing models to determine feed composition.

In Third World countries, the formulated feed is less common than in the United States or the European Union, but their feed industries are growing rapidly. As developing countries achieve economic growth and become more urban, they are adopting more intensive livestock and poultry production methods. Improved feed milling will provide better balanced rations that meet the specific nutritional requirements of livestock and poultry.

Forage Uses

Maize is an excellent forage for cattle, water buffalo, and other large ruminant animals. When the entire maize plant is harvested and utilized as a feedstuff, it surpasses all other crops in average yield of dry matter and of digestible nutrients per hectare. Maize is commonly fed to livestock as fodder, stover, or silage.

Fodder consists of the entire fresh, cured, or dried plants. Maize is an important fodder source in South and Southeast Asia and in Egypt. In Asia, especially in the higher elevations, farmers plant maize at high densities (above 120,000 plants per hectare) and then progressively thin green plants from the field to feed to their livestock. By the time flowering starts, farmers have eliminated some 50,000 plants. The rest are left to form ears and reach physiological maturity. In Egypt, farmers progressively strip lower maize leaves from the plant and feed them as roughage to their livestock.

Stover includes the dried stalk minus the ears. It is used by many maize farmers in developing countries as a roughage feed for livestock. Maize stover contains 30 to 40 percent of the plant's total nitrogen, three-fourths of its potassium, sulfur, and magnesium, and almost all of its calcium.

Whole plant maize silage, consists of the entire plant, cut, chopped, and placed in a structure where anaerobic fermentation allows it to be stored safely. Silage maize is harvested 2 to 3 weeks earlier than maize harvested only for grain. It is an excellent feedstuff for dairy and beef cattle. Silage maize is an important feed in temperate areas such as the United States, Canada, and Europe, where it makes up 10 to 30 percent of the area is planted to maize. Silage maize is also important in temperate South America, Mexico, Turkey, and parts of east Asia, especially China and North Korea. Maize cob mix—silage made from maize ears—is increasingly popular as a swine feed in Europe.

Wet Milling

Maize is attractive for the manufacture of starch and sweeteners by the wet milling process (Watson 1988). It produces abundant starch (65% of kernel weight) that is easily recovered (95% of all present) and that can be processed to high purity (99%). Byproducts generated from the starch-making process also have significant economic value.

From its beginning in the 19th century in the United States, the maize wet milling industry has developed sophisticated technology to separate the components of the maize kernel by chemical and physical means into a large number of useful products. Starch is the main product of the wet-milling process and is used either in starch products or is further processed into other industrial food and nonfood products.

In the USA, 75 percent of the starch produced by wet milling is hydrolyzed products, mainly sweeteners (National Corn Growers Association 1992). The rest is industrial starch. The wet-milling industry also produces various modified maize starches by for paper lamination, textile warp-sizing coatings, and laundry finishing. Other modified starches are used in the food industry as a gel to add body to various processed foods.

Wet millers desire maize kernels that are free from foreign matter, have low moisture content, are physically sound, and that yield high percentages of starch, oil, and protein. High moisture content may lead to aflatoxin problems, it forces the mill to handle more total weight, and it can upset the water balance in the milling system. The primary consequence of processing damaged maize kernels is loss of oil, a valuable byproduct.

The wet-milling process begins by soaking clean maize grain in water, thus swelling the kernels. During the soaking or steeping process, nutrients are absorbed by the water (steepwater). When steeping is complete, the steepwater is drawn off and condensed. During the subsequent milling process, the maize germ is separated from the kernel. The germ is further processed to remove the oil, and the remaining germ meal is isolated for later use. After the starch has been removed, the rest of the maize kernel, containing starch, gluten, and bran, is screened and the bran is removed. The remaining mixture of starch and gluten is separated by centrifugal action. The starch portion is dried, or modified and dried, and sold to food, paper, and textile industries, or is further converted to various sweeteners.

Modified a few steps further, maize starch becomes a glucose syrup or a dried crystalline glucose powder. Glucose is the major source of energy in human and animal nutrition. Liquid glucose (corn syrup) and dried crystalline glucose help to sweeten many foods in the United States. Three-fourths of the processed foods in the supermarket contain some corn syrup or dextrose. The largest volume of this glucose goes into confections, followed by baking uses, and beer brewing. By using an enzyme to isomerize glucose to fructose, the wet-milling industry has also developed a high-fructose corn syrup (HFCS) that is as sweet as cane sugar, but with a differing taste to some users. Maize sweeteners made up more than half the U.S. sweetener (nondiet) market in 1991, a five-fold increase since 1970 (National Corn Growers Association 1992).

First introduced in the late 1960s, HFCS comes in several forms with multiple applications, offering many of the functions of glucose syrups, plus greater sweetness. HFCS-42, consisting of 42 percent fructose (levulose) and 58 percent glucose, goes into jams, jellies, bakery and canned products, and condiments. HFCS-55, with 55 percent fructose and the same sweetness as sucrose, was specifically developed as a sweetener for soft drinks. Today it dominates the U.S. soft-drink market. Crystalline fructose, which has a sweetness 1.8 times that of sucrose, has recently entered the U.S. market.

HFCS has made dramatic inroads in food-processing industries in the developed economies, and has become the pre-eminent maize sweetener in beverages. HFCS is now beginning to enter the marketplace in the developing world, mainly in the newly industrialized developing countries. To date, however, HFCS is only sold in tank loads and requires large capital investments for equipment.

In wet milling, for every 100 kilograms of maize processed into starches, sweeteners, and ethanol, about 30 to 35 kilograms of byproduct, mainly protein and fiber, are left over after the recovery of starch and oil. The most economically important byproducts are animal feed products. Twenty-five to thirty kilograms of feed products are produced: mainly maize gluten meal, maize gluten feed (from the bran), maize germ meal, and maize steepwater (condensed fermented extractions). The steepwater also provides a culture medium in which antibiotics such as penicillin and aureomycin can be produced. Maize oil is produced from the germ and is used in cooking oil, salad oil, margarine, where it is appreciated for its clean flavor, ability to withstand high temperatures, and its high level of polysaturates. For every 100 kilograms of grain processed, 2 to 3 kilograms of maize oil is produced.

Starch chemistry is still an infant science, with each new revelation leading to new products and uses. One super-absorbent starch combination, perfected by a U.S. Department of Agriculture research team, is capable of soaking up 2,000 times its own weight in moisture. Called "Super Slurper," it has quickly found uses among makers of diapers, bed-pad protectors, and fuel filters. A new frontier is developing in the packaging industry for cornstarch-based biodegradable plastics for manufacturing goods such as cutlery, cups, and plates.

Maize germplasm varies in its starch make-up, especially in the proportion of the long-chained starch molecule called amylose that is present in the kernel. Normally, the kernel contains about 30 percent amylose. Breeders, however, have been able to develop maize strains with up to 80 percent amylose. Amylose starch can make an edible film suitable for wrapping foodstuffs. Foods wrapped in edible film can be cooked without unwrapping—there is no paper or plastic to throw away. This product is being used to encapsulate products from chemicals to birdseed.

Dry Milling

A wide range of products for food, feed, and industrial uses are produced from maize by dry milling. In the United States, the dry milling industry consumes about 2 percent of maize production, or 4.2 million tons (National Corn Growers Assocation 1992). Uses of dry milled maize products are animal feed (31%), brewing (23%), breakfast cereals (22%), other food (16%), and industrial products (8%).

Three basic processes are used: (1) tempering-degerming process, (2) stone-ground or non-degerming process, and (3) alkaline-cooked process. The most common is tempering-degerming. After cleaning and drying, the maize is tempered to 20 percent moisture, and the majority of the pericarp and germ is removed, leaving the endosperm. The bulk of the endosperm, known as the tail hominy fraction, proceeds through the degerminator and is dried, cooled, and sifted. A portion of this fraction is isolated as large flaking grits, which are used for products such as breakfast cereals. Further separation produces smaller grits, meals, and flours. The bran and germ fractions are passed through another part of the degerminator, and the germ is separated from any remaining endosperm. This process yields crude maize oil, hominy feed, bran products, standard meal, and prime grits, meals, and flours.

In the non-degerming process, the entire kernel is ground intact, producing an oily flour. White grain is generally used, and the primary products are hominy grits and cornmeal. These products are essentially whole-ground maize with very little of the pericarp and germ removed. The non-degerming process improves the flavor and nutritional value but shortens the time the meal stays fresh.

In the alkaline-cooked process, the maize is boiled in a lime solution for 5 to 50 minutes, depending on intended use, and then steeped for 2 to 12 hours. The cooked and steeped maize is washed to remove excess alkali and loose pericarp tissue, and the resulting maize product is ground into flour, which, mixed with water, forms the dough used in the production of tortillas, maize chips, and other snack items. These maize products can be stored at room temperature without becoming rancid (in contrast to whole-kernel flour, which becomes rancid with time).

Dry millers require large kernels that are physically sound and have less than 15 percent moisture and high test weights (weight per unit volume). About 65 percent of the maize emerges as prime products and 35 percent as byproducts. Millers use yellow dent maize to produce most products, although white maize is often preferred for meals and flours.

Fermentation and Distilling

Various alcoholic beverages and industrial products are produced by maize distilling and fermentation industries. Malt converts maize starch to sugar, which is fermented by yeast to ethyl alcohol and carbon dioxide. The byproducts that remain including oil, vitamin and protein concentrates, and grain-germ mixtures are used as a livestock feed.

Maize is increasingly being used as the carbohydrate source in beer-making, either as a dry adjunct (mainly dry milled maize flakes, grits) or as a liquid adjunct (mainly maize syrups and dextrose). Maize is also used to manufacture distilled, grain-neutral liquors, such as whiskey and vodka. The fermentability of maize starches and sweeteners has also made maize an important feedstock for ethanol (ethyl alcohol). Ethanol is being used both as a complete fuel substitute in gasoline engines or in a mixture of 10 percent ethanol and 90 percent gasoline, called gasohol. In the United States in 1991 about 10 million tons of maize were fermented for ethanol production. However, without government subsidies, ethanol is not competitive with petroleum-based fuels.

Composite Flours

The idea of using composite flours to supplement wheat flour for making bread and biscuits is quite old. Increased global wheat production since the Green Revolution and a reduction in the global wheat prices in real terms, has boosted wheat consumption in many tropical countries where the climate precludes growing wheat that has good bread-making quality. For some time, such countries depended on imported wheat or wheat flour received as food aid or purchased from wheat surplus countries. Many of these tropical countries are now hard pressed for foreign exchange and are restricting imports of wheat and flour and use of foreign exchange for this purpose.

Milling and baking researchers have shown that it is technically feasible to substitute flours of crops like maize, sorghum, millet, or cassava for wheat flour to a limited extent. Most of this research has focused on technical feasibility of making such composite flours. However, the economics of substitutions and effects on the industry have not been closely examined.

Several countries in sub-Saharan Africa that have little wheat production capacity and rising demand for wheat bread have a potential market for composite flours. Yet composite flours are used commercially only in Zambia (6% maize flour) and Zimbabwe (10% maize flour). In Latin America several countries have conducted research on composite flours, but only Brazil uses composite flour (made with cassava and maize).

The only tropical region that has exhibited little interest in composite flours is Southeast Asia. This may be due to the fact that, with a few exceptions, per capita wheat and bread consumption is low and the primary consumers are medium to high income groups who expect bread of very high quality.

Substitution of nonwheat flour is limited to a maximum of 10 to 20 percent, otherwise the quality of bread falls to an unacceptable level. Biscuits can tolerate up to 30 percent substitution in composite flours. Generally flours made from cassava and sorghum have been found to be superior to maize flour for substitution in bread and biscuits. However, in many countries, the supply of sorghum and cassava of the appropriate quality is more uncertain.

Work done at the International Maize and Wheat Improvement Center and elsewhere has shown the feasibility of substituting maize flour for wheat flour up to 10 percent without any appreciable difference in the quality of composite bread. For biscuits the substitution could be much higher—even more than 30 percent. Hard endosperm quality protein maize (QPM) (see Chapter 7) has opened new possibilities for wider use of this type of maize in composite flours.

Researchers at the University of Arizona have found that when cornmeal from hard-endosperm QPM (with suitable genetic modifiers) is used with wheat flour in bread making, the dough has better quality than mixtures containing cornmeal made from normal maize. The QPM cornmeal gives the dough a better balance of sulfur-containing amino acids.

In Brazil, EMBRAPA scientists have reported the superiority of QPM (particularly the variety BR 451, which is under commercial cultivation in Brazil) over normal maize varieties in preparation of composite maize-wheat flours for bread making. Composite flours are already being used in commercial bread bakeries in Brazil.

Ghana. Photo copyright © Tony Freeth. Used by permission.

3

Maize Production Environments and Methods in the Developing World

Principal Maize Environments

Maize environments in developing countries can be grouped into four major classifications: tropical, subtropical, temperate, and highland (Table 3.1). These classifications do not correspond precisely to geographic definitions of the tropics, subtropics, and temperate regions (Fig. 3.1). Rather they are based on agroclimatic criteria: (1) minimum and maximum mean temperatures during the growing season, (2) altitude above sea level, and (3) to a lesser extent, latitude (Table 3.2).

Between the equator and 34° north or south, altitude has a major effect on temperature regimes. Moving from lowland to highland elevations, maize environments change from tropical to subtropical to cool highland elevations. Lowland maize areas (below 1,000 meters elevation) between 23° and 34° north and south, though geographically in the

TABLE 3.1 Major maize environments in the developing world (000 ha).

Region	Tropical	Subtropical	Temperate	Highland
Southern Cone, South America[a]	8,700	4,000	1,800	0
Andean region, South America[b]	1,600	300	0	500
Mexico, Central America & Caribbean	4,300	1,600	0	3,000
West & Central Africa	5,200	300	0	50
East & Southern Africa	2,100	7,000	0	1,500
North Africa & Mideast	0	1,000	1,100	0
South Asia	5,600	1,400	0	600
Southeast Asia	8,700	200	0	0
East Asia	500	1,300	19,300	550
Total	36,700	17,100	22,200	6,200

[a] Argentina, Brazil, Chile, Paraguay, Uruguay.
[b] Colombia, Ecuador, Peru, Bolivia.
Source: FAO Production Yearbooks; CIMMYT Maize Program 1988.

35

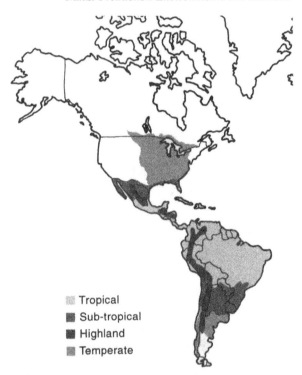

FIGURE 3.1 Major maize environments.

subtropics, are classified as having a tropical environment in the summer season. In temperate environments (latitudes above 34°), considerably more hours of daylight occur during the maize growing season than in the tropical and subtropical latitudes, and photoperiod-insensitive genotypes are required. In highland environments, which have the coolest growing-season temperatures, the growing season is long and frost is a danger at the beginning and end of the season.

In tropical and subtropical environments, particularly, a variety of abiotic stresses depress maize yields. Occasional moisture stress characterizes about 20 million hectares of maize in developing countries (24% of total area) during the growing season, and moisture stress is common on another 10 million hectares.

Problem soils that are low in organic material and have limited crop productivity unless fertilizer is applied occupy 30 million hectares of arable land in developing countries, mainly in tropical environments. Approximately 9 million hectares of this area have soils that are acidic, and many of these soils have toxic levels of free aluminum that inhibit root development.

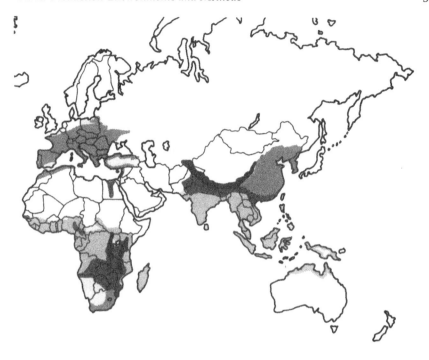

Maize has one of the highest photosynthetic rates of all the food crops. But when maize is grown under low light intensity, its rate of photosynthesis drops significantly, and so consequently does its yield potential. Maize yield potential in cloudy monsoon climates, like those of Asia from Pakistan to southern China, is considerably less than the potential in areas that have less cloud cover and more hours of bright sunlight, such as most temperate environments in the South America and north China.

The amount of water the maize crop requires varies according to

TABLE 3.2 Agroclimatic criteria used in maize environmental classifications.

Environment	Mean growing season temperature (°C)			Altitude (m)	Latitude N - S
	Min.	Max.	Avg		
Tropical	22	32	28	< 1,000	33° or lower
Subtropical	17	32	25	< 1,600	23° to 34°
	17	32	25	1,000-1,800	23° or lower
Temperate	14	24	20	< 500	higher than 34°
Highland	7	24	16	> 1,800	23° or lower
	9	25	18	> 1,600	23° to 34°

temperature, humidity, wind, and soil fertility. Ninety-nine percent of the water taken up by the plant is lost through transpiration. Transpiration rates, therefore, determine water demand. These rates are lower with cooler temperatures and higher with warmer temperatures. Under high temperatures, as well as other conditions conducive to rapid transpiration rates such as low humidity and wind, water losses from the maize plant may exceed the uptake of water, causing wilting.

Temperatures also affect energy accumulation in the maize plant. High temperatures cause the maize plant to grow faster and burn off much of the energy (carbohydrates) that it accumulates through photosynthesis. Above about 45°C, the net assimilation rate (NAR) becomes zero because respiration exceeds photosynthesis. Low temperatures, on the other hand, retard maize plant growth and development. Below about 10°C, the NAR also becomes negative because photosynthesis essentially stops. Further, in highland and temperate environments, early or late-planted maize crops are also vulnerable to plant-killing frosts.

The type and severity of the biotic stresses—diseases, insects, and weeds—are functions of many factors, including geographic location, natural vegetation, climate and weather variations, crop rotations, and tillage practices. Some disease and insect species thrive in hot, humid tropical environments; others find cool, humid climates more hospitable. And some are indigenous to specific geographic regions—their migration to other continents is checked by oceans and major land barriers such as deserts or mountain ranges. Many diseases, insects, and weeds are parasitic to maize plant, and are capable of killing the plant or severely depressing grain and forage yields.

Production stresses tend to be greater in hot, humid tropical environments than in cooler, less humid subtropical and temperate environments. To date, plant breeders have had more success in developing genotypes that have resistance to important diseases than ones that have resistance to the major insect pests. Where improved maize genotypes are used, insect pests generally are more of a threat than diseases unless insecticides are applied.

Tropical Environments

Maize-growing areas with environments classified as tropical cover 36.7 million hectares, 45 percent of the total maize area in the developing world (Table 3.1) and are found in all developing country regions, except North Africa and the Middle East (Fig. 3.1). These environments are characterized by a high mean temperature (around 28°C) during the growing season (Table 3.2). They have a high mean maximum temperature (around 32°C) and a high mean minimum temperature (around 22°C). Most of the tropical maize area (32 million ha) is located between the equator and 23° north or south in about 60 countries of Asia, Latin America, and Africa. The remaining 4.7 million hectares of tropical environments are located in what technically are the subtropical latitudes—23° to 33° north and south—at elevations below 1,000 meters, where high day-time and night-time temperatures occur throughout the growing season, e.g., in Paraguay and northern Argentina, the Punjab of India and Pakistan, and the southern-most provinces of China. (However, these areas are considered to have subtropical climates if maize can be grown during the winter or spring seasons.)

Moisture Availability

In most tropical environments, maize requires 600 to 700 millimeters of moisture that is well distributed over the growing season. Nearly 70 percent of the tropical maize area—25 million hectares (including irrigated areas)—usually has sufficient moisture for maize production. In the remaining tropical maize environments, the crop is frequently stressed by drought on 7 million hectares and occasionally stressed on another 4.7 million hectares. Only 2 million hectares of the tropical maize area are irrigated, mainly in Asia. In Latin America and sub-Saharan Africa, less than 5 percent of the maize area is irrigated.

Grain and Plant Maturity Preferences

Farmers in tropical environments grow both white and yellow maize of dent and flint grain types. White grain maize is generally preferred as a human food; yellow grain is preferred as an animal feedstuff. Yellow maize is grown on 20 million hectares in the tropical environments (60% of the total area), mostly in South and Southeast Asia and in the Southern Cone of South America. Genotypes with white flint and dent grain are grown on 15 million hectares in the tropical environments throughout most of sub-Saharan Africa, lowland Mexico, Central America, and the Caribbean.

Intermediate- to late-maturing genotypes (110- to 120-day maturity) are grown on 70 percent of the tropical maize area and early maturing genotypes (90- to 100-day maturity) are planted on the rest. Early maturity genotypes are preferred where double and triple cropping is practiced or where the growing season is short because of limited rainfall. As cropping intensity increases in the more favored production environments, the demand for high-yielding maize varieties that have early maturity has increased.

Diseases and Insects

Maize in tropical environments is attacked by an array of plant diseases that can cause significant damage (Table 3.3). The most pervasive are southern rust, southern leaf blight, leaf spot, and several types of stalk and ear rots. In addition, several diseases are of regional importance. Various forms of downy mildew attack maize in Southeast Asia and Central and Western Africa. Maize streak virus, transmitted by a leafhopper, also severely lowers yields in some seasons in tropical Africa. Corn stunt disease, also transmitted by a leafhopper, can cause economic losses in Mexico, Central America, and parts of South America. The parasitic weed striga also causes costly losses in tropical Africa and in parts of India and Southeast Asia.

Estimates of the potential and actual area currently affected by important maize diseases are shown in Table 3.4. So far, downy mildew has not gained a wide foothold in Latin America, and streak virus has been confined to sub-Saharan Africa. However, if either of these diseases crosses ocean barriers, substantial economic losses could result.

The maize crop can be seriously damaged by insect pests in tropical environments (Table 3.5). The most significant are several classes of maize borers and armyworms. Termites cause significant crop damage in many areas in sub-Saharan Africa. Cutworms and rootworms can also cause serious economic losses in Southeast Asia. Grain weevils cause significant grain damage throughout tropical environments. Recently the larger grain borer, inadvertently imported from the Americas, is spreading across West Africa, and has the potential of causing enormous losses to stored grain. In addition, several leafhoppers are vectors for maize streak virus in Africa and for maize stunt disease.

Subtropical Environments

Subtropical environments are a function of altitude and latitude. They are found from the equator to 34° north and south and account for 17 million hectares or 21 percent of the maize area in developing countries. Subtropical environments have cooler mean temperatures (around 25°C) and longer growing seasons than tropical environments (Table 3.2). High mean day-time temperatures in subtropical environments are similar to those in tropical areas. But because subtropical environments occur at higher elevations or in higher latitudes, they have lower mean night-time temperatures plus a distinctly cooler winter season. They are found in the mid-elevation areas in tropical latitudes of eastern and southern Africa, southern Brazil, Mexico, and Central America, and in the subtropical latitudes of Brazil, Egypt, South Asia, and China.

TABLE 3.3 Important diseases in major maize environments.

Common name	Causal organism	Affected regions
Tropical		
Southern rust	*Puccinia polysora*	All
Southern leaf blight	*Helminthosporium maydis*	All
Leaf spot	*Curvularia* spp.	All
Stalk rots	*Fusarium* spp., *Diplodia* spp.	All
Ear rot	*Fusarium* spp., *Diplodia* spp., *Aspergillus flavus, Cladosporium* spp.	All
Downy mildew	*Perenosclerospora* spp.	Southeast Asia, Sub-Saharan Africa, Central America, lowland Mexico
Stunt	Corn stunt spiroplasma	Mexico, Central America
Streak virus	Maize streak virus	Sub-Saharan Africa
Striga (witchweed)	*Striga* spp.	Sub-Saharan Africa, Asia
Tarspot	*Phyllachora maydis*	Americas
Subtropical		
Northern leaf blight	*Exerohilum turcicum*	All
Common rust	*Puccinia sorghi*	All
Ear rot	*Fusarium* spp., *Diplodia* spp., *Aspergillus*	All
Stalk rot	*Fusarium* spp., *Diplodia* spp., *Pythium* spp., *Erwina cartovara**	All
Late wilt	*Cephalosporium maydis*	North Africa, South Asia
Tar spot	*Phyllachora maydis*	Latin America
Streak virus	Maize streak virus	Africa
Downy mildew	*Perenosclerospora* spp.	West and Central Africa
Striga	*Striga* spp.	Sub-Saharan Africa, South Asia
Temperate		
Northern leaf blight	*Exerohilum turcicum*	All
Southern leaf blight	*Helminthosporium maydis*	All
Maize dwarf mosaic	Maize dwarf mosaic virus	East Asia
Head smut	*Sphacelotheca reiliana*	East Asia
Stalk rot	*Fusarium* spp.	All
Ear rot	*Fusarium* spp., *Diplodia* spp., *Aspergillus flavus*	All
Highland		
Common rust	*Puccinia sorghi*	All
Northern leaf blight	*Exerohilum turcicum*	Warmer areas
Stalk rot	*Fusarium* spp., *Diplodia* spp.	All
Ear rot	*Fusarium* spp.	All
	Diplodia spp.	Warmer areas

* Bacteria.
Source: CIMMYT.

Moisture Availability

About 90 percent of the maize area in subtropical environments is rainfed. Of the 2 million hectares that are irrigated, about half is in Egypt and the remainder in India and Pakistan. Moisture is generally adequate on 9 million hectares (includes irrigated land), 47 percent of the area of subtropical environments. About 8 million hectares of subtropical maize lands experience occasional moisture stress, and another 3 million hectares experience frequent to usual moisture stress.

Grain and Plant Maturity Preferences

White maizes (flints and dents) are grown on about 11 million hectares in subtropical maize environments (57% of total) in developing countries, mostly in East Africa, southern Africa, Mexico, and East Asia. Yellow maizes (mainly flint grain types) are grown on 8 million hectares, mostly in South Asia and the Southern Cone of South America.

Late-maturing genotypes (160 days) are grown on 10 million hectares, intermediate-maturing genotypes (130 days) on 7 million hectares, and early maturing genotypes (100 days) on about 2.5 million hectares. Most of the demand for early maturing germplasm is in South Asia, where double and triple cropping is practiced.

Diseases and Insects

Northern leaf blight, common rust, and stalk and ear rots (caused by *Fusarium* and *Diplodia* spp.) are the predominant diseases in subtropical environments (Table 3.3). Maize streak virus is a problem in East and

TABLE 3.4 Estimates of tropical maize area affected by diseases causing economic losses.

Disease	Area	Potential area affected (million ha)	Maize area affected* (%)
Downy mildew	East/southern Africa	2	50
	Southeast Asia	9	100
	South Asia	3	30
	Latin America	15	10
Streak virus	East/southern Africa	2	60
	West/Central Africa	5	90
	Asia	13	0
	Latin America	15	0
Southern rust	Worldwide	37	60
Southern leaf blight	Worldwide	37	25
Ear rots	Worldwide	129	20

* Require disease-resistant germplasm.
Source: CIMMYT maize program data and authors' estimates.

TABLE 3.5 Major insect species causing economic losses in maize.

Common name	Scientific name	Affected regions
Tropical environments		
Stem borers		
Spotted stem borer	*Chilo partellus*	Asia, East Africa
Oriental corn borer (Asian maize borer)	*Ostrinia furnacalis*	Asia
Lesser cornstalk borer	*Elasmopalpus lignosellus*	Americas
Pink stem borer	*Sesamia cretica*	Africa
African maize borer	*Sesamia calamistis*	Africa
African maize stalk borer	*Busseola fusca*	Africa
African sugarcane borer	*Eldana saccharina*	Africa
Sugarcane borer	*Diatraea saccharalis*	Americas
Neotropical corn borer	*Diatraea lineolata*	Central and South America
African leafhopper	*Cicadulina* spp.	Africa
Fall armyworm	*Spodoptera frugiperda*	Americas
Cutworm	*Agrotis* spp.	All regions
Termites	*Microtermes* spp.	Africa, Asia
Grain weevils	*Sitophilus* spp.	All regions
Larger grain borer	*Prostephanus truncatus*	Latin America, parts of Africa
Subtropical environments		
Stem borers		
European maize borer	*Ostrinia nubilalis*	North Africa, Mideast
Oriental corn borer (Asian maize borer)	*Ostrinia furnacalis*	Asia
Spotted stem borer	*Chilo partellus*	Africa
African maize stalk borer	*Busseola fusca*	Africa
Sugarcane borer	*Eldana saccharina, Diatraea saccharalis*	Africa
Southwestern corn borer	*Diatraea grandiosella*	Americas
Fall armyworm	*Spodoptera frugiperda*	Americas
Corn earworm	*Heliothis zea, H. armigera*	Americas
Termites	*Microtermes* spp.	Africa, Asia
Grain weevils	*Sitophilus* spp.	All
Temperate environments		
Southwestern corn borer	*Diatraea grandiosella*	Southern Cone, South America
Lesser corn stalk borer	*Elasmopalpus lignosellus*	Southern Cone, South America
Oriental corn borer	*Ostrinia furnacalis*	East Asia
European corn borer	*Ostrinia nubilalis*	North Africa, Mideast
Fall armyworm	*Spodoptera frugiperda*	Southern Cone, South America
Corn earworm	*Heliothis zea*	Southern Cone, South America
Cutworms	*Agrotis* spp.	All
Highland environments		
Corn earworm	*Heliothis zea*	Americas
Cutworms	*Agrotis* spp.	All
Grain weevils	*Sitophilus* spp.	All

Southern Africa. Late wilt is a problem in Egypt. Striga, a parasitic weed, causes economic losses in subtropical environments of East and Southern Africa.

Several classes of borers and maize earworms are economically important in maize in subtropical environments (Table 3.5). Stem borers, armyworms, earworms, rootworms, and storage insects cause the greatest losses.

Temperate Environments

Temperate maize environments are found above 34° latitude north and south and account for 22.3 million hectares, 27 percent of the total maize area in the developing world (Table 3.1). During the growing season, these environments have a mean temperatures of 20°C and receive more hours of solar radiation per day than do latitudes closer to the equator. The mean maximum temperature during the day, 24°C, and the mean minimum temperature during the night, 14°C, are ideal for maize growth and development. However, frost is a threat during the spring and fall. In the developing world, China has about 87 percent of the temperate maize-growing area and Argentina has 8 percent.

Moisture Availability

Maize, which is rainfed in 95 percent of the temperate environment, requires 400 to 500 millimeters of rainfall during the growing season. About 65 percent of the total area normally receives sufficient moisture. In the rest of the temperate maize area—mainly in parts of China, North Africa, and the Middle East—the maize crop experiences occasional to frequent moisture stress.

Grain and Plant Maturity Preferences

Nearly all the maize grown in temperate environments has yellow grain, either dent or flint. On 75 percent of the area, farmers prefer genotypes with intermediate to late maturity (130 to 160 days). Parts of China require early maturing genotypes (110-day maturity) to fit maize into relay and double-cropping systems.

Diseases and Insects

Diseases in temperate environments (Table 3.3), generally cause less economic damage to maize than diseases in warmer environments, especially because improved, disease-resistant varieties and hybrids are widely planted. However, northern leaf blight, southern leaf blight, ear rot, maize dwarf mosaic virus, and stalk rots can cause economic losses in these environments unless resistant genotypes are grown or some other form of disease control is employed.

Insects can cause economic losses in maize in temperate environments (Table 3.5). Stem borers, armyworms, earworms, and cutworms are the most serious pests. Their damage is greatest in temperate environments with warmer climates.

Highland Environments

Highland environments account for about 6.2 million hectares or 8 percent of the developing country maize area (Table 3.1). These environments are found between 1,800 and 3,000 meters elevation in the tropical latitudes (equator to 23° north or south) and between 1,600 and 2,700 meters in subtropical latitudes (23° to 34° north and south). Mean daily temperatures are 16° to 18°C, with the mean maximum temperature during the day about 25°C and the mean minimum temperature during the night about 8°C. Cool temperatures early in the crop cycle often inhibit plant growth, and frosts threaten the crop during the spring and fall. Highland maize-growing environments are found in Mexico (45% of total area), eastern and southern Africa (25% of total area), Central America, the Himalayan areas of Pakistan, Nepal, India, and China, and the Andean region of South America.

Moisture Availability

Virtually all highland maize is rainfed. Moisture is generally adequate in the highlands of Central America and southern Mexico, in East Africa, and in the Asian countries bordering the Himalayas. In contrast, maize grown in highland areas of central Mexico and the Andean countries of South America—which account for about 50 percent of the total highland area—experiences frequent to severe moisture stress.

Grain and Plant Maturity Preferences

Most of the maize grown in highland environments has white grain of varying textures and hardness. Varieties that have white grain and semi-flint and semi-dent grain types are grown in highland Mexico, the Himalayas, China, and sub-Saharan Africa and account for 75 percent of highland maize. Yellow flints are grown in the highlands of Guatemala and parts of the Himalayan area and account for 20 percent of highland maize. Floury and *morocho* maizes, which have soft, large kernels of various colors (white, yellow, gray, black), predominate in the Andean highlands and account for about 5 percent of the total highland maize area.

The length of the maturity period in maize is related to elevation because as the mean temperature drops, the maize-growing season becomes markedly longer. Between the equator and 30° north or south, a full-season maize variety generally matures in 125 days at 1,600 meters elevation; it takes 160 days to mature at 2,000 meters elevation; and it takes 220 days to mature at 2,500 meters elevation. At elevations between 2,700 and 3,000 meters, a full-season variety can take 330 to 360 days to mature.

Early maturing varieties are grown on about 60 percent of the highland area, especially in locations between 23° and 33° north and south, where moisture availability and the threat of frost limit the growing season. In highland environments where moisture and temperature are not limiting, farmers want late-maturing germplasm.

Diseases and Insects

Important maize diseases that can cause economic losses in highland environments include northern leaf blight, common rust, tarspot, and stalk and ear rots (Table 3.3).

The insects that cause the most pervasive economic losses in highland areas are maize earworm, cutworms and rootworms, and grain weevils (Table 3.5). Maize earworms are perhaps the most damaging insect pest because they also set the path for ear rot to infect the maize ear.

Cropping Patterns

In developing countries, maize is grown as a solid-stand monocrop, in relays and rotations with other crops over sequential growing seasons, and intercropped in association with other crops during the same growing season. Important maize-based cropping patterns are shown in Table 3.6. Probably 60 percent of the total maize in the developing world is grown in various multiple-cropping patterns (patterns in which more than one crop is grown on the same land in a year).

Sequential Cropping

Maize is grown as a solid stand in sequential cropping patterns on about 40 percent of the total maize area in the developing world, generally in rotation with other crops but sometimes as a monoculture i.e., maize is planted one season per year, year after year. Such patterns are primarily functions of biology and economics. As temperatures and moisture become more limiting, cropping pattern tends to shift from monoculture to sequential cropping where maize is grown in rotation with other crops. Also, when farm sizes are larger and maize production is mechanized and more agricultural chemicals are applied, the crop is grown as a solid stand, usually in rotation with other crops.

In Southeast Asia and the Pacific, maize is often grown in rotation with rice. In irrigated areas of South Asia, maize has traditionally been grown as a summer crop with supplemental irrigation, frequently in rotation with wheat. Recently, however, maize is being increasingly planted in the winter and spring. In many areas, maize is replacing late-planted wheat. Maize is higher yielding and more profitable and fits better in rotation with the main summer crops, principally late-maturing rice and cotton.

In Argentina and Chile, maize is grown almost exclusively as a solid-stand crop within a 3-year maize-wheat-soybean rotation. In Mexico and Central America, commercial maize producers tend to grow maize as a solid-stand crop. Where climate and season permit, maize may ei-

TABLE 3.6 Important maize-based cropping patterns.

Region	Environment	Mono-cropped (%)	Major rotation	Inter-cropped (%)	Major associations
Southern Cone, South America	Temperate	100	wheat, soybeans	0	none
Andean region, South America	Highland	20	maize, quinoa, Lupinus	80	beans, squash, potatoes
Central America	Tropical	70	cotton, sugarcane	30	beans, sorghum
	Highland	20	maize, wheat	80	beans, squash
West & Central Africa	Tropical	20	maize, cowpea	80	cassava, cocoyam, groundnut, cowpea
East & Southern Africa	Subtropical	30	maize	70	beans, cowpea groundnut, squash
	Highland	30	maize, potatoes	70	squash, beans, peas
North Africa	Subtropical irrigated	80	wheat, berseem	20	chickpea
South Asia	Tropical/ subtropical	80	wheat, rice, potato	20	chickpea, potato
	Highland	70	maize	30	squash, beans
Southeast Asia	Tropical	80	rice	20	soybean, groundnut, mungbean
East Asia	Temperate	70	wheat, rice	30	wheat, mungbean
	Highland	50	wheat, rice	50	mungbean
	Tropical/ subtropical	50	wheat, peas, barley	50	potato

Sources: Authors' estimates; American Society of Agronomy 1978

ther be grown in a double-cropping pattern or as a monoculture, in which it is planted season after season on the same land.

Relay Cropping

With the development of improved early maturing crop varieties and the spread of irrigation, two to three crops can be grown annually where moisture is sufficient and climate permits. Relay cropping is a form of multiple cropping in which two or more crops are grown simultaneously during part of their life cycles. A second crop is planted after the first crop has reached its reproductive stage of growth but before physiological maturity.

Relay cropping involving maize is expanding in many parts of the developing world. In north China, maize frequently is planted as a relay crop between the rows of winter wheat 1 month or less before the wheat harvest. (Where the growing season is longer, maize is planted after wheat, sequentially.) In southern China and northern Vietnam, maize is grown as a relay crop after rice. In this system, maize seedlings are raised in special nurseries for 3 weeks and then transplanted into harvested rice fields. In Vietnam, farmers transplant more than 150,000 hectares of winter maize into paddy fields to fit a rice-rice-maize rotation.

In the tropical environments of West Africa, some farmers have adopted a relay-cropping system involving maize and cowpeas. They plant an early cowpea variety (60-day maturity) at the beginning of the rainy season. After about 35 days, they plant maize between the cowpea rows. By the time the maize plants emerge and begin the active vegetative growth stage, the cowpeas have reached physiological maturity and are harvested.

Intercropping

Intercropping is defined as growing two or more crops simultaneously on the same field. Several types of maize intercropping systems exist. Mixed intercropping involves various arrangements between maize and other crop species in certain proportions. Row intercropping involves growing maize and one or more additional crops simultaneously in alternating rows. Strip intercropping involves growing maize and one or more additional crops simultaneously in strips wide enough to permit cultivation of each crop independently, but narrow enough for the crops to interact agronomically.

When farmers select combinations of crops and planting schedules that minimize interspecific competition for available light, water, and nutrients, intercropping systems on small land holdings (1 to 2 hectares) can produce greater total output per hectare than when the individual crops are grown separately. As farm size becomes larger, intercropping becomes less practical because mechanized production systems are generally required, and planting, harvesting, and application of inputs become more difficult.

Intercropping is more common among near-subsistence, small-scale farmers, where traction sources are largely human or animal and where temperature and moisture availability permit cultivation during most of the year. In areas with 400 to 700 millimeters of rainfall, simultaneous cropping tends to be practiced with crops of similar maturity. In areas with 700 to 1,000 millimeters of rainfall, the species in crop mixtures commonly have different maturities, so that the later maturing crops can make maximum use of moisture at the end of the season. In areas with rainfall above 1,000 millimeters, simultaneous intercropping is practiced in combination with sequential multiple cropping.

Low-density intercropping is employed by many small-scale farmers in areas subject to numerous biotic and abiotic stresses to help ensure a dependable family food supply. Such complex crop mixtures are more dynamic biologically than a solid stand crop and are more likely to yield adequately despite adversities such as insect attack and drought. In West Africa, for example, cowpea growers who do not have access to insecticides intercrop cowpeas with maize to deter the insects that attack cowpeas (by spreading out the distribution of cowpea plants).

In highland Latin America among small-scale producers, about 80 percent of all maize is grown in association with other crops, such as climbing or bush beans (*Phaseolus* spp.) and various squashes. In sub-Saharan Africa, 70 percent of the maize area is usually intercropped. In West and Central Africa, maize is grown in association with starchy staples such as cassava, cocoyam, or plantain or with grain legumes such as cowpeas or groundnuts. In East and southern Africa, maize is commonly intercropped with beans, cowpeas, groundnuts, pumpkins, or pigeonpeas in lowland areas and with squash, beans, potatoes, peas, and rapeseed in highland areas. In Asia, highland maize is also frequently intercropped with soybeans, mungbeans, field beans, squash, potatoes, peas, or groundnuts.

Examples of row and strip farming include the row intercropping of maize and sorghum in Central America and the strip intercropping of maize and annual tree crops, called alley farming, in parts of Southeast Asia and the Pacific, especially on several Indonesian islands. In this system, maize is grown in strips that alternate with strips of nitrogen-fixing trees and shrubs and trees planted as hedgerows. Properly managed, these nitrogen-fixing shrubs can annually add 20 to 30 kg/ha of nitrogen to the soil, help to control erosion, and serve as a source of firewood for home cooking.

Production Technologies

Maize is an important crop for 70 million farm families around the world, 80 percent of whom live in developing countries. Vast differences exist in the technology employed to produce maize (Table 3.7).

There are striking differences in the labor farmers in different countries use to produce a ton of maize. The level of mechanization, itself a function of farm size and farmers' financial resources, has a major effect on labor productivity. The use of seed, fertilizers, and improved cultural practices also makes an important difference in labor productivity.

In the United States, a typical maize farmer plants 60 hectares of maize, uses a high-yield technology, and manages a highly mechanized operation requiring a two-person team. The average yield, 7.5 t/ha, gives a 450-ton harvest.

In France, the average maize farmer plants about 30 hectares of maize and has a highly mechanized operation, requiring a two-person work force. High levels of fertilizer and hybrid seed are used. The average yield, 6.7 t/ha, gives a harvest of about 200 tons.

In Argentina, a typical farmer plants 30 hectares. He relies on mechanized production systems (though not as intensive as in the United States or Europe) and requires a three-person work force. The farmer uses a hybrid, but little or no fertilizer, relying on the natural soil richness of the Pampas. The average yield is about 3.6 t/ha, giving a total harvest of 108 tons.

TABLE 3.7 Comparison of maize production technologies in selected countries.

Country	1989-91 yield (t/ha)	Estimated labor[a] (days/t)	Use of		
			Mechanization	Fertilizer	Improved seed
USA	7.5	1.3	High	High	High
France	6.7	1.7	High	High	High
Argentina	3.6	4	High	Low	High
Turkey	4.0	12	Inter.	Inter.	Inter.
China	4.3	18	Inter.	High	High
Egypt	5.7[b]	22	Inter.	High	Inter.
Mexico	1.9	35	Inter.	Inter.	Low
Guatemala	2.0	37	Low	Inter.	Inter.
Kenya	1.7	50	Low	Inter.	High
Pakistan	1.4	65	Low	Inter.	Low
Ghana	1.4	65	Low	Low	Inter.
Haiti	0.8	75	Low	Low	Nil

[a] Average labor (person-days) employed by typical maize farmer to produce 1 ton of maize, based on average farm size, yield, and labor input in each country.

[b] Irrigated maize production requiring additional labor for water management.

Source: Yield data from 1991 FAO Production Yearbook; labor productivity measurements calculated by authors.

In China, a typical maize farmer plants 2 hectares, possibly hiring a tractor for land preparation, requires a two-person work force, and employs a high-input production technology. The average yield would be 4.3 t/ha with a harvest total of 8.6 tons of grain.

In Kenya, a typical maize farmer plants 2 hectares, requires a two-person work force, and uses improved seed, probably hybrid, but relatively small amounts of fertilizer. The average yield is 1.7 t/ha, giving a total grain harvest of 3.4 tons.

In Ghana, a typical maize farmer plants 0.75 hectares, requires a two-person work force, and uses an improved open-pollinated variety based upon improved germplasm, but little or no fertilizer. The average yield would be 1.4 t/ha, giving a total harvest of 1.05 tons.

Small-scale farmers currently relying on human labor and hand tools and low-yielding production technologies have several opportunities to increase productivity. First, where animal traction can be used for land preparation and weed control, a two-person work force can manage 4 to 6 hectares instead of 1 to 2 hectares. Second, by using improved maize genotypes, fertilizers, and more optimum cultural practices, current yields can often be doubled or tripled.

(Note: In this chapter, all data on environments, grain types, and moisture, soil, disease, and insect stresses are from CIMMYT Maize Program 1988.)

Mexico. Photo courtesy of CIMMYT.

4

Maize Research Systems in the Developing World

Introduction

Maize is the most thoroughly researched of all the cereals, and the one in which the greatest productivity advances have been achieved. Plant breeding advances have been complemented by similar advances in soil chemistry and physics, plant physiology, plant pathology, entomology, cytogenetics, biometrics, and molecular biology. These various research findings and products have been synthesized into new, higher yielding technologies that have been extended to farmers in many maize-growing areas. The result has been a succession of productivity-enhancing maize technologies that have propelled maize to its current position as the cereal with the world's highest average yield.

The Beginnings of Maize Research

Until the 19th century, maize improvement was in the hands of farmers who selected the seed from preferred plant types of landraces or populations for subsequent sowing. By the early 1800s, farmer-breeders in the United States had identified the sexual nature of maize and had observed the beneficial effects that could be achieved when certain strains of maize were inter-mated. With this knowledge, a number of progressive farmer-seedsmen in North America began crossing different strains of maize, and developing and selling seeds from superior varieties based on individual plant selections. The variety Reid Yellow Dent, which remains the basis of most U.S. Corn Belt hybrids today, was a selection developed by John Reid, an Illinois farmer, in 1856.

The groundwork for modern genetic improvement of maize and other crops was laid by Darwin in his writings on the variation of species (published in 1859) and by Mendel through his discovery of the laws of inheritance (reported in 1865). The rediscovery in 1900 of Mendel's laws provoked tremendous interest and research in plant genetics. Maize, because of its agricultural importance, its high level of genetic heterogeneity and heterozygosity, and the ease with which self- and cross-pollination could be accomplished, quickly became a favorite species for study by U.S. plant geneticists.

The institutional foundations for modern maize research were laid in the United States during latter half of the 19th century and early part of the 20th century. In 1862, Congress established the U.S. Department of Agriculture (USDA) and the publicly supported land-grant colleges of agricultural and mechanical arts in every state. In 1887, Congress passed a law that provided for the establishment of experimental research stations at each land-grant college, as well as closer collaboration between colleges and the USDA. In 1914, the final key organization, the Extension Service, was set up as the education arm of the USDA research service and the state agricultural experiment stations and charged with introducing new technologies and providing technical services to farmers and ranchers.

The U.S. Hybrid Maize Revolution

Darwin's observations about hybrid vigor inspired W. J. Beal, an agricultural botanist at Michigan State College, to carry out in 1877 the first controlled crosses between maize varieties for the sole purpose of increasing yields through hybrid vigor. Although Beal did not use the term hybrid vigor, his results proved the existence and importance of this phenomenon. The rediscovery of Mendel's laws at the turn of the century led George Shull, a botanist in New York, to develop inbreeding as a technique in maize improvement. He grasped the idea that, once produced, inbred strains had to be crossed to be of any practical usefulness. In papers published from 1908 to 1911, Shull coined the word heterosis, which is commonly used to mean hybrid vigor.

However, it was one of Shull's contemporaries, Edward M. East, who first saw the significance of the inbred-hybrid method of varietal development in maize. East developed a number of high-yielding single-cross hybrids that had a 25 to 30 percent yield superiority over the best open-pollinated varieties then available. But although the hybrid crosses were often quite vigorous, the inbreds themselves were small, miserably weak, unattractive, and unproductive, and they made single-cross hybrid seed production too costly. The solution to this problem was achieved by Donald F. Jones, a former student of East, who developed a technique that made hybrid maize seed production practical. In 1918, Jones and his research associate, Paul Mangelsdorf, published a paper describing the double-cross hybrid, which used two single-cross hybrids as the parents, rather than two inbred lines, for the production of commercial seed. This technique overcame the problem of the low seed yield of inbred lines and was a giant step toward reducing the cost of hybrid seed production.

With the double-cross method for hybrid seed production available, the stage was set for the rapid replacement of varieties by hybrids. By the late 1920s, some markedly superior hybrids were available for commercial use, and in 1926, Henry A. Wallace, who later became U.S. Secretary of Agriculture and U.S. Vice President under Franklin Roosevelt, founded Pioneer Hi-bred International, the first seed company organized solely to produce hybrid seed.

Publicly funded maize breeders at the state agricultural universities and USDA research institutions developed most of the new inbred lines and commercial hybrids, but private companies were responsible for seed production. By the mid-1940s, there were thousands of small- to medium-scale seed growers (mostly family farm operations) producing hybrid maize seed. The diffusion of hybrid maize technology was initially delayed by the Great Depression (1929 to 1933). In 1934, only 0.4 percent of the maize area in the United States was planted to hybrids; 10 years later, hybrids covered 90 percent of the Iowa Corn Belt and 60 percent of the total U.S. maize area.

Following World War II, cheap nitrogen fertilizer became available in the United States, and farmers began to use significant amounts of chemical fertilizers. It soon became apparent that most of the hybrids were not suited to the increased soil fertility. They responded to application of nitrogen fertilizer by developing weak stalks susceptible to stalk rot and lodging. By 1955, a few hybrids (double crosses and three-way crosses) reached the market that had been especially bred for higher planting density and higher soil fertility. With these management-responsive hybrids and improved fertility conditions, U.S. maize yields made a quantum jump during the 1960s followed by another jump during the 1970s, as a new generation of higher yielding single-cross hybrids became available.

The fact that hybrids have closed pedigrees (which are protection against being copied) and must be regenerated each year from the inbred parental crosses to avoid loss of vigor in the F_2 generation provided the opportunity for profit that encouraged the establishment of hybrid seed companies. Over time, profits from hybrid seed sales have been poured back into research, which has produced continuing expansion in yield potential.

Attempts to Transfer U.S. Hybrid Technology Abroad

The success of the hybrid maize revolution in the United States in the late 1930s and early 1940s provoked considerable interest among farmers and agricultural scientists in other countries. The Great Depression, and then World War II, however, interrupted the exchange of U.S. maize inbred lines and hybrids and scientific information with other nations. After the cessation of hostilities, there was an urgent need to expand food production in war-torn countries in the shortest time possible.

With its well-developed agricultural research and extension systems, and high-yield food production technologies, the United States was in the position to lead international efforts to introduce high-yielding crop technologies. Because of its dual food and feed uses, and the existence of high-yielding hybrids, maize was identified as a priority crop for introduction in food-deficit areas of Europe and Asia.

During the 1950s, USDA, U.S. foreign aid programs, and FAO distributed a large number of U.S. F_1 double-cross hybrids to maize researchers in Europe, Asia, Latin America, and Africa (McMullen 1987). The performance of the U.S. hybrids in other environments met with mixed results. In temperate areas with similar climatic conditions, such as in southern Europe, Chile, and China, the U.S. hybrids performed well and were quickly accepted by farmers once seed production and distribution systems were established. In northern Europe, however, genotypes with earlier maturity and greater cold tolerance were required. For these colder environments, French national researchers and FAO scientists, working with U.S. germplasm and local genetic resources, were successful in developing early maturing, cold-tolerant hybrids by the late 1950s. These new materials made maize a competitive crop in the northern latitudes and caused Europe's total maize yields and production to more than double between 1960 and 1990.

Attempts to introduce U.S. hybrids and germplasm in subtropical and tropical environments met with far less success. U.S. hybrids were sent for testing in India, Pakistan, Brazil, North Africa, and the Middle East, but failed to gain a foothold because they lacked adaptation to the climate, diseases, and insect pests in those countries. The inbred lines in the double crosses were generally too weak to withstand the difficult production environments of tropical areas. In addition, the yellow dent grain of the U.S. hybrids was not popular in many countries. Nevertheless, U.S. hybrids and inbred lines were extensively used as breeding stocks in many tropical and subtropical maize improvement programs to take advantage of various traits. Soon, it became clear that U.S. hybrid maize technology was not suited to the production conditions in most developing countries, particularly those with subtropical and tropical environments.

However, most developing countries had little capacity to carry out maize improvement research. Governments neglected food crop research because the dominant economic development theories at the time held that developing countries should bypass their traditional agricultural sectors, except for cash crops, and they should focus on industrial growth, with agriculture providing surplus labor as needed to develop the industrial and urban sectors. Although the populations of developing countries were mostly rural, development funds were channeled into industrial investments—generally parastatal—rather than the modernization of the agricultural sector.

The Rockefeller Foundation's Pioneering Role in Maize Research

The Rockefeller Foundation had an enormous influence on the development of maize research systems in developing countries. The Rockefeller Foundation began its first international agricultural research effort in 1943, in collaboration with the government of Mexico. The goal was to improve productivity and increase production of the nation's basic grains, with maize being the primary staple. At that time, Mexico had few trained maize scientists. Mexican farmers grew traditional varieties and used little or no fertilizer to maintain soil fertility; national maize yields averaged about 500 kg/ha; and national food deficits were growing each year.

The Mexican-Rockefeller Foundation program set out to develop improved food crop technologies (mainly varieties and crop management practices) in the shortest time possible to help Mexico regain food self-sufficiency and to establish a national research infrastructure capable of providing a continuing stream of improved crop technologies to farmers. Over a 20-year period, hundreds of Mexican researchers received Rockefeller scholarships to pursue graduate school training, often at U.S. universities. The foundation also financed the construction and equipping of many agricultural research stations and support laboratories. And, it helped to launch a national agricultural research institute and a postgraduate agricultural college.

In 1948, the Rockefeller Foundation established an agricultural research program in Colombia, structured along the lines of the Mexican program. The Colombia-Rockefeller Foundation maize improvement program was able to draw on the germplasm development work under way in Mexico.

The maize improvement research program sponsored by the Rockefeller Foundation and Mexico had international dimensions almost from the start. It provided in-service training opportunities to other scientists from Latin America and, later, Asia, and it freely shared its research information and improved germplasm with scientists outside Mexico. These activities helped to pioneer the concept of international networking in agricultural research, which has greatly accelerated progress in plant breeding research worldwide.

One of the first tasks undertaken by maize scientists in the Rockefeller Foundation programs in Mexico and Colombia was to begin collecting native maize races to serve as a germplasm base for improvement. These efforts were later augmented by the U.S. National Academy of Sciences, which formed the Committee on Preservation of Indigenous Strains of Maize to sponsor genetic resource collection and classifications projects throughout the Americas.

During the 1950s, the Rockefeller Foundation expanded its international maize research assistance beyond Latin America to Asia. In 1957, the Rockefeller Foundation-Indian maize improvement program was launched and made rapid progress in developing improved germplasm. Breeding progress was accelerated by the availability and introduction of improved germplasm complexes provided by the Mexican and Colombian maize programs.

At the end of the 1950s, the general condition of national agricultural research in the developing world was feeble. Dozens of nations, many just emerging from colonial pasts, had little or no organized maize research under way. Developing nation had few trained maize researchers or technicians. Consequently, the fame and impacts of the Rockefeller Foundation's bilateral agricultural research projects in Latin America and Asia led many governments to request similar cooperative research programs, which could not be met on a bilateral basis.

To cope with the demand, the Rockefeller Foundation agricultural program began to seek ways to development a permanent international agricultural research system and began experimenting with other organizational models. One step in this process was the establishment of regional maize research programs and networks.

In 1954, Rockefeller Foundation established the Central American Maize Program, which involved five countries. This program later evolved into an independent regional crops research association, called the PCCMCA, whose members now include most food crop scientists in Central America, Panama, and Caribbean. The PCCMCA operates regional maize variety yield trials and sponsors an annual scientific conference that is widely attended by national and international crop researchers.

To tie together—and provide follow-up to—the various bilateral maize research programs it operated, the Rockefeller Foundation established the Inter-American Maize Program for Latin America and posted staff in Mexico, Colombia, and Chile. It also established the Inter-Asian Corn Program in 1964, with Rockefeller Foundation maize scientists first posted to India and, after 1966, to Thailand.

The International Maize Research System

Although the regional maize research networks supported by the Rockefeller Foundation facilitated the exchange of germplasm and scientific information, they lacked the institutional base for many research and training activities. To help accelerate the generation and transfer of improved technology to Third World farmers and to strengthen national agricultural research systems as quickly as possible, the Rockefeller Foundation, joined by the Ford Foundation, set out to establish a network of commodity-specific international agricultural research centers, well-equipped, -staffed, and -financed, with independent professional management structures, and located in the major food-producing ecologies of the Third World.

In 1960, the two private foundations were prime movers in launching the first full-fledged international agricultural research center, the International Rice Research Institute (IRRI), in cooperation with the government of the Philippines. Shortly thereafter, the two foundations helped to establish the International Maize and Wheat Improvement Center (CIMMYT) in Mexico in 1963 (officially incorporated in 1966), the International Center for Tropical Agriculture in Colombia in 1967, and the International Institute for Tropical Agriculture (IITA) in Nigeria in 1967.

The Green Revolution fueled by the CIMMYT and IRRI wheat and rice technologies led to a considerable expansion in the number of international research centers. In 1971 the World Bank, the United Nations Development Programme, and FAO established the Consultative Group on International Agricultural Research (CGIAR) to marshal financial support for a strategic research network focused on the major food crops and production issues in the developing world. By the 1990s, the CGIAR consisted of about 40 donors, government aid agencies, international and regional organizations, and private foundations, and it was providing $280 million to operate a network of 17 international agricultural research centers scattered around the globe.

Maize improvement has been a major research priority since the inception of the network of international agricultural research centers. Among the CGIAR centers, CIMMYT has a worldwide mandate for maize improvement in the tropics and subtropics and mounts the largest international maize research program. Maize improvement at IITA, on the other hand, focuses on western and central Africa. In addition, the International Plant Genetic Resource Institute is involved in collection and preservation of ancient maize races.

CIMMYT and IITA have overlapping research interests in sub-Saharan Africa. To avoid duplication, the centers have worked to improve research coordination and collaboration. Both institutions freely exchange breeding materials and facilitate the international testing of each others' more advanced materials. During the 1980s, the centers operated a joint program to introduce maize streak virus resistance into CIMMYT's most widely used tropical germplasm complexes and into several dozen improved open-pollinated varieties based on this germplasm complex. In the 1990s, by mutual agreement, IITA has taken the lead in maize germplasm development and crop management research support in western and central Africa, while CIMMYT has assumed similar responsibilities for East Africa and southern Africa. Since 1991, the CGIAR system, and consequently CIMMYT and IITA, has faced serious budget constraints, which has led to significant staff reductions and the contraction of maize research, training, and technical assistance.

CIMMYT's Maize Program

CIMMYT's maize program originated nearly 30 years ago with a considerable carryover of personnel, breeding materials, and organizational philosophy from the former Mexican-Rockefeller Foundation maize research program. With its new international mandate, the maize program took on new staff and expanded the scope of the breeding materials and the germplasm development objectives.

One of CIMMYT's first steps was to develop a network of experiment stations in Mexico located in different maize-growing environments where maize improvement work could be carried out with precision. With this accomplished, scientists focused during the early 1970s on the formation and initial stages of improvement of a broad range of germplasm complexes, primarily for tropical maize-growing environments. After several cycles of improvement these maize germplasm complexes were tested widely in probable zones of adaptation throughout the developing world. Although many of these improved maize materials had excellent yield potential, most were still too tall and leafy and too variable in several agronomic characteristics to gain widespread acceptance among farmers. Overcoming these deficiencies became the principal task of the CIMMYT maize breeders, in collaboration with national program scientists in developing countries.

During the 1970s, the CIMMYT maize program underwent a significant reorganization. A key feature of the new organization was international maize testing program designed to serve as a combined germplasm development and delivery mechanism. Through international progeny testing trials, national collaborators participated directly in the improvement of populations and the development of experimental varieties. CIMMYT has emphasized the development of maize gene pools and advanced populations. Until the mid-1980s, open-pollinated varieties were the end product of its breeding work. Since then, development of inbred lines for hybrid combinations has received much greater priority.

CIMMYT maize improvement priorities include higher yield potential combined with greater yield dependability. Significant aspects of this research program are described in Chapters 6 and 7.

CIMMYT has also offered a broad range of training programs, mainly in Mexico, to maize researchers from national maize programs. Training courses have been offered in maize improvement (breeding, pathology, entomology), maize agronomy and crop management, cereal quality laboratory techniques, and production economics. Several thousand developing-country maize scientists are alumni of these in-service courses. More individually tailored and specialized training courses have been offered to several thousand senior maize researchers from developing countries.

CIMMYT has also given considerable technical assistance to national programs. Coordinated by CIMMYT staff posted to regional programs, the assistance has included consultation services by maize program scientists, collaborative research projects with national programs, scientific information services, and workshops and conferences. In its earlier years, CIMMYT assigned scientists to work within some national maize programs for fixed terms.

IITA's Maize Research Program

IITA's maize improvement program was initiated in the early 1970s. It built upon the work of several predecessor organizations working in West Africa. The first maize research was initiated by the Colonial West Africa Maize Department at its station, Moor Plantation, in Ibadan, Nigeria, during the 1950s, and centered on developing germplasm with resistance to southern rust and southern leaf blight.

Following Nigerian independence, the Rockefeller and Ford foundations financed a bilateral maize research program in Nigeria until IITA was established in 1967. The best breeding materials from the Rockefeller Foundation-supported national maize research programs in Latin America and Asia, and later from CIMMYT, were supplied to IITA.

IITA's maize program has sought to develop high-yielding germplasm that has increased yield dependability under the agroclimatic stresses found in humid and subhumid tropical and mid-elevation production environments primarily located in western and central Africa (see Chapters 6 and 7). In particular, IITA work to develop high-yielding open-pollinated varieties and hybrids with resistance to maize streak virus has been outstanding.

In crop management research, IITA has been active in studying intercropping systems involving root crops and legumes in which maize is an important component. IITA scientists also are seeking new crop management systems to maintain soil fertility (e.g., alley cropping), control various diseases and insect pests (e.g., biological control and integrated pest management), and provide food security when droughts occur.

IITA trains maize researchers from African nations. In-service training courses are routinely offered in maize improvement and crop management. IITA also provides fellowships for African researchers to complete MS and Ph.D. theses, working under the supervision of IITA scientists and using the institute's research facilities. Finally, visiting scientist fellowships are routinely awarded to senior maize scientists from other research institutions.

Other International Maize Research Programs

Aside from the international agricultural research centers, other public-sector organizations operate international maize research programs. During the 1950s and 1960s, FAO scientists pioneered maize improvement research in many developing countries. They distributed superior germplasm to national scientists and helped organize and conduct maize improvement programs. As one of the co-sponsors of the CGIAR, FAO discontinued most of its maize research in the 1960s, after maize research programs were under way at CIMMYT and IITA.

Funded largely by their national foreign aid programs, some agricultural universities and government research institutes in developed countries have also been active in maize improvement research in developing countries. Hundreds of researchers from developing countries have received master's and doctoral degrees at agricultural universities in North America, Europe, and the Soviet Union. Scientists from U.S. universities and USDA research stations also have been involved maize improvement research activities in developing countries. Outstanding early examples of such collaboration were the U.S.-funded maize improvement work of American scientists in Tiquisate, Guatemala, during the 1950s and in Kitale, Kenya, during the 1960s.

Other developed maize-growing countries have also funded international agricultural research programs. The most notable have been France, through the Institute for Tropical Agricultural Research (IRAT), which conducts research in Africa and the Caribbean in former French colonies, and the Maize Research Institute of Yugoslavia, which has provided technical experts and training assistance to help strengthen national maize research programs in the Middle East, South Asia, and southern Africa.

Several regional research organizations have engaged in maize research, such as the Inter-American Institute of Agricultural Sciences (IICA) and the Centro de Agricultura Tropical para Investigacion y Enseñanza, both headquartered in Costa Rica. Some research network associations also fund maize research. Examples are Safgrad (Sahelian countries of Africa), Saccar in Southern Africa, Prociandino (Andes Region, South America), and Conasur (Southern Cone, South America).

International Dimensions of the U.S. Maize Research System

Although most maize improvement research in the United States during the past 75 years has been publicly funded, private research predominates today. Since the 1960s, private maize research expenditures in the United States have increased 20-fold and now account for two-thirds of the total maize research. Probably 70 to 80 private companies are engaged in maize breeding research. Another 20 to 25 private companies, not traditionally in the seed business, are engaged in biotechnology research to improve maize.

In 1990, the U.S. hybrid maize seed industry generated over US$1.2 billion in sales. Ten hybrid maize seed companies in the United States account for about 60 percent of national hybrid seed sales. The rest, however, comes from several hundred small- and medium-sized seed companies, often family enterprises, that produce either public hybrids or those of another private seed company for a fee.

Despite the significant rise in private maize research, publicly funded maize research still plays an important complementary role to private research. More than a dozen state agricultural universities and experiment stations and USDA have maize research programs that have made distinguished research achievements. Public inbred lines and hybrids are still widely used by private seed companies, both for direct seed production and as breeding materials to develop improved versions. As the private sector has increased its expenditures in maize research, the public sector has tended to move "upstream" in its maize research activities, developing improved breeding techniques and new germplasm complexes to serve as source materials for various characteristics. Private research organizations have focused on utilizing the information and germplasm complexes generated in public research programs to develop finished hybrids.

Several leading transnational companies based in the United States and Europe are now becoming more active in seed production in the developing world, either directly or through joint venture activities with local companies.

National Maize Research Systems

National agricultural research systems in developing countries vary considerably in their organization models, orientation, and development stages. They have been created as an instrument of the state to promote agricultural development. The rationale for the creation and development of publicly funded national research systems, however, has also reflected a practical reality. Since most developing countries were poor and had little or no research structure, the state appeared to be the only means to generate the necessary organization and financing to develop and transfer improved technology to farmers. Likewise, the state was seen as the only entity able to carry out other agricultural sector activities such as input supply, marketing, credit, and provision of machinery.

Public-sector Maize Research

Today, perhaps as many as 70 developing countries conduct some types of publicly funded maize research (57 developing countries grow 100,000 hectares of maize, or more). These national maize research programs differ in size, scope, and scientific capabilities. A CIMMYT survey of national maize research operating budgets in 44 developing countries in 1990 shows that countries in sub-Saharan Africa spend relatively more on research per hectare of maize than do other regions (Table 4.1). This probably reflects higher inherent research costs due to poorly developed infrastructure and the absence of complementary private-sector research. However the higher relative investments in sub-Saharan Africa (and Latin America) may also reflect the influence of IITA and CIMMYT on national maize research programs in these regions.

As Table 4.2 indicates, there is a great disparity in the numbers of maize scientists in different national maize research systems in the developing world. China, Brazil, Mexico, and India, the four largest

TABLE 4.1 Annual public-sector investment in maize research, 1989-90.

Region	Average maize program budget[a] (US$)	Expenditures relative to	
		Maize area (US$/000 ha)	Researchers (US$/person)
Sub-Saharan Africa	197,000	289	15,000
North Africa & West Asia	99,000	178	3,000
South, East, & Southeast Asia	435,000	115	4,000
Latin America	439,000	213	16,000
Avg, 44 countries	305,000	176	8,000

[a]Budget data unavailable for Costa Rica, Paraguay, and Zambia.
Source: CIMMYT 1992.

maize-producers, which account for two-thirds of total production in the developing world, have 140 to 600 maize scientists each, who are primarily engaged in plant breeding and crop management research. Relative to the area of maize grown, Venezuela and Egypt have many

TABLE 4.2 Estimated number of public and private maize researchers in selected countries, 1990.

Country	1989-91 maize area (million ha)	Maize researchers (no.)		Total maize researchers (no./million ha)
		Public sector	Private sector	
Asia				
China	20.8	633	0	30
India	5.9	168	50	37
Philippines	3.7	43	117	32
Indonesia	3.0	37	7	14
Thailand	1.6	22[a]	65	53
Pakistan	0.9	26	13	46
Latin America				
Brazil	12.1	59	252	21
Mexico	7.0	85[a]	55	20
Colombia	0.8	10[a]	19	36
Guatemala	0.6	16[a]	22	26
Venezuela	0.5	24	25	103
North Africa & West Asia				
Egypt	0.9	69	23	107
Turkey	0.5	19	24	83
Africa				
Tanzania	1.8	14	2	9
Nigeria	1.6	28[a]	11	25
Kenya	1.5	58[a]	2	39
Malawi	1.3	8	2	8
Zimbabwe	1.1	6[a]	19	22
Ethiopia	1.1	12	0	11
Mozambique	1.0	10	11	21

[a] Excluding maize staff from CGIAR centers.
Source: CIMMYT 1992.

researchers. In part, the explanation is that the countries maintain large public-sector programs despite the increasing amount of private maize research activity.

Another 15 countries, accounting for about 20 percent of Third World maize production, had an average of 40 scientists engaged in maize research; the next 20 countries, accounting for about 10 percent of production, had an average of 15 scientists engaged in maize research; finally, about 20 small countries, accounting for 5 percent of developing country maize production, had an average of 6 scientists engaged in maize research.

In 1980 the CGIAR estimated that national maize research systems in 26 developing countries were spending 0.3 percent of the value of maize production. Applying this spending ratio to the total value of developing country maize production in 1980, suggests that approximately US$50 million a year was being spent on maize research. At the same time, the international centers, principally CIMMYT and IITA, were spending another $9 million on maize-related research (salaries, administrative costs, and overhead), while the private sector in developing countries was probably spending $6 million. Thus, in total, $65 million a year was being spent on maize-related research directed at the developing world.

In 1990, if national maize research systems were still spending 0.3 percent of the value of maize production, that would be equal to $63 million. In 1990, the CGIAR international centers were spending another $12 million on maize-related research. However, the biggest increase has come in the private sector in developing countries, which is probably spending $20 to $25 million in maize-related research. Over the decade, then, maize-related research directed at the developing world increased by nearly 50 percent, in nominal terms, to a total of $95 to $100 million.

Most publicly funded maize research is undertaken by national and provincial agricultural research institutes. Maize research programs in most national agricultural institutes tend to emphasize plant breeding. Maize agronomy and crop management research are frequently the responsibility of other research departments and programs.

Some developing countries such as Brazil, Pakistan, and Nepal also have established national or provincial multidisciplinary maize research institutes. Countries with mixed federal-provincial research systems such as India, Pakistan, and Brazil have established national agricultural research councils to guide the work of provincial and federal research institutions. In such systems, nationally coordinated commodity research programs were established for major crops such as maize.

Maize research in most developing countries began in the agricultural colleges, which had the greatest concentrations of scientific personnel. Over time, however, most universities ceded maize improvement to research institutes and departments supported by ministries of agriculture or ministries of science and technology. After the creation of national agricultural research institutions during the 1960s and 1970s, most public research funds flowed to these organizations. Starved for research funds, the agricultural universities became chiefly teaching institutions.

Exceptions to this pattern are developing countries that established universities along the lines of the U.S. agricultural universities, which have teaching, research, and technology-transfer functions. In Asia, the maize breeding programs of the University of the Philippines and Kasetsart University (Thailand) have produced outstanding new germplasm complexes and varieties. Several of India's state agricultural universities also operate substantial maize research programs. In Latin America, the National Agricultural University of Peru, the University of Sao Paulo in Brazil, and the National University of Chile have active maize improvement programs.

Once a nation's private sector becomes a major factor in maize breeding, public-sector organizations tend to focus more on the germplasm needs of farmers in marginal production areas, often emphasizing open-pollinated varieties and informal seed production and distribution systems. Public organizations also step back and work more on development of new basic germplasm complexes as breeding sources, rather than in trying to compete with the private sector in the development of hybrids. Table 4.3 show the importance of universities and other public and private organizations in national maize research systems.

Private-sector Maize Research

The modernization of agriculture implies an increased role for purchased inputs and services and lends itself to greater role for certain types of private research and development. In the United States, private research primarily focuses on crop protection (insecticides, herbicides, fungicides), plant breeding (hybrid maize development, especially), and food processing. Two general types of private organizations undertake maize research. One is the agricultural input company that engages in various aspects of maize research, such as plant breeding, plant pathology, entomology, soil fertility, and cereal technology to develop products and services to sell directly to the farmer. The second type is the corporate farmer organization (cooperatives, commodity associations) whose members tax themselves to support certain research activities (plant breeding, crop management) designed to serve their needs.

Until recently, the private sector has had a minor role in maize research in developing countries, except for market-oriented economies with temperate regions such as Argentina, Chile, and South Africa. In the 1980s, however, private maize breeding research expanded significantly in the developing world, especially Latin America and Asia. Further increases in private investments in maize research and hybrid seed production are likely in the future.

TABLE 4.3 Composition of national maize research systems.

Country	Public research institutions	Agricultural universities	Private seed companies
South America			
Argentina	●●	●	●●●
Bolivia	●●●		
Brazil	●●	●	●●●
Chile	●	●	●●●
Colombia	●●●	●	●
Ecuador	●●		
Paraguay	●●		
Peru	●●	●●	●●
Venezuela	●	●●	
Mexico, Central America, and Caribbean			
Costa Rica	●	●	
Cuba	●●		
Dominican Rep.	●●		
El Salvador	●●		●●
Guatemala	●●		●●
Haiti	●		
Honduras	●●		
Mexico	●●	●	●●
Nicaragua	●●		
Panama	●	●	
East and Southern Africa			
Ethiopia		●●	●
Kenya	●●	●	●
Malawi	●●		●●
Mozambique		●●	
Somalia	●●		
South Africa	●	●	●●
Tanzania	●●		
Uganda	●		
Zambia	●●		●
Zimbabwe	●●	●	●●

Continued.

One reason is that the largest unexploited markets in which to sell improved maize seed, and other agricultural inputs and services, are in the subtropical and tropical environments, where most farmers still employ low-yielding traditional production practices. Most of the maize area in the industrialized countries is already planted to high-yielding hybrids, and there is little scope to expand. But in the developing countries, roughly half of the maize area—40 million hectares—is planted to traditional or only marginally improved varieties. Thus, there is enormous potential to increase the area planted to high-yielding genotypes.

Table 4.3. Continued.

Country	Public research institutions	Agricultural universities	Private seed companies
West and Central Africa			
Benin	●	●	
Burkina Faso	●●		
Cameroon	●●		●
Central African Rep.	●		
Côte d'Ivoire	●●		●
Ghana	●●		
Nigeria	●●	●	●●
Senegal	●●		
Togo	●●		
Zaire	●●		
North Africa and the Middle East			
Egypt	●●	●	●●
Morocco	●		
Turkey	●		●●
South Asia			
Bangladesh	●		
India	●●	●●	●●
Nepal	●●		
Pakistan	●●		●
Southeast Asia			
Burma	●●		
Indonesia	●●	●	
Philippines	●●	●●	●
Thailand	●	●●	●
Vietnam	●●		
East Asia			
China	●●	●	
North Korea	●●●		
South Korea	●●	●	●

●●● = dominant. ●● = strong. ● = limited.
Source: Authors' estimates.

Second, thanks to the maize research undertaken during the past two decades by international agricultural research centers and national agricultural research systems, the inventory of profitable technological components—improved genotypes and production practices—available to extend to farmers is considerable. Often it only requires minor adaptive research to be suitable for introduction and distribution among farmers.

Third, popular disenchantment with the ineffectiveness of public input-supply organizations has led governments to radically reform old development policies. Increasingly, governments are turning many functions over to the private sector and forcing the remaining public input-supply organizations to compete without preferential subsidies.

Private seed companies generally develop their seed production and distribution systems before engaging in much proprietary plant breeding research. Initially, most private seed companies rely on inbred lines and hybrids developed by other public institutions. However, as seed sales develop and profits grow, a portion of the profits is plowed back into research to develop proprietary inbred lines and hybrids that can give the company a sales advantage in the commercial marketplace. Most of the new private maize breeding programs—transnational and national—draw heavily on the improved germplasm complexes developed by strong national maize improvement programs and by CIMMYT and IITA.

In the past two decades, the large maize seed companies have become increasingly active overseas in seed production and distribution. By the late 1980s, transnational hybrid maize seed companies exported more than 50,000 tons of hybrid maize seed annually, worth $50 million and sufficient to plant 2.5 million hectares (Pray and Echeverria 1991). Initially, these maize seed companies were only interested in countries with temperate production environments, such as Argentina, Chile, Turkey, and southern Europe, where U.S. hybrids were already fairly well adapted. Over time, these companies have expanded their local maize research activities to remain competitive.

Transnational seed companies are responsible for most of the private maize breeding research under way in the tropics and subtropics. Currently they are operating maize breeding programs in more than 25 developing countries. In addition, private national seed companies are also engaged in maize plant breeding research in countries where government policy fosters private research.

Profiles of National Maize Research Systems

Following are brief profiles of national maize research systems in various regions of the developing world. The national research systems of Ghana, Zimbabwe, Thailand, China, Guatemala, and Brazil are described in detail in Chapter 10.

Latin America and Caribbean. In the 1950s and 1960s, South American governments established semi-autonomous national institutions to carry out agricultural research and technology transfer activities in which maize was invariably a top priority crop. Argentina established a semi-autonomous national agricultural research institute in 1956, followed by Peru in 1960, Mexico and Ecuador in 1961, Colombia in 1962, and Chile in 1964. In 1973, Brazil created a completely new federal agricultural research corporation to support and complement the work of state research institutes. These institutional developments resulted in greatly increased maize research activity during the 1960s and 1970s.

The trend in national maize research systems in South America during the 1980s has been toward a mixed system involving both public and private initiatives. Publicly funded national maize research programs still account for the majority of maize research expenditures in most countries, although budgets have declined during the 1980s. In Argentina and Chile, national maize research is now almost exclusively carried out by private organizations. In countries that still have large numbers of small-scale, resource-poor farmers, e.g., the Andean countries, Venezuela, Brazil, Paraguay, publicly funded maize research programs still account for most of the scientific activity.

Maize research in Mexico, Central America, and the Caribbean has traditionally been carried out primarily by publicly funded national research institutes and programs. In some countries, agricultural universities have also mounted substantial maize improvement programs. CIMMYT, located in Mexico, has maize breeding programs under way for tropical, subtropical, and highland production environments.

East and Southern Africa. The oldest maize research program in Africa was established in Rhodesia (Zimbabwe) in 1932. Initially, breeders developed several improved open-pollinated varieties from local germplasm and several varieties from the Americas. Since 1950, the public-sector program has been totally oriented to hybrids, and seed production has been undertaken by the Zimbabwe Seed Co-op. Over time, public research in Zimbabwe has declined and private research has increased (see Chapter 10).

Maize research in East Africa began in the early 1950s under the colonial East African Agriculture and Forestry Research Organization (EAAFRO), which operated in Kenya, Tanzania, and Uganda. A maize research section was organized at Kitale, Kenya, and began operation in 1955. In 1963, the U.S. Agency for International Development financed a major cereals research project, based in Kenya, in which USDA scientists were assigned to develop national maize breeding programs in East

Africa. With the improved national breeding stocks and breeding methodologies, Kenya began releasing open-pollinated varieties and hybrids, which then were effectively produced and distributed through the Kenya Seed Company. Most farmers in more favorable production environments, including small-scale growers, use the company's hybrids.

Today, relatively strong maize breeding research programs—and vigorous maize seed sectors—exist in Kenya, South Africa, Zambia, and Zimbabwe, as indicated by the high proportion of the total maize area planted with hybrids. Publicly funded national maize research programs predominate in most countries, the major exceptions being in Zimbabwe and South Africa where private-sector research is more important.

West and Central Africa. Maize research began in West Africa during the colonial period as a result of a devastating rust epidemic. Southern rust was accidentally introduced into West Africa in 1949 and within 10 years had reached epidemic proportions in much of tropical Africa. Scientists from the colonial West African Research Department in Nigeria and from EAAFRO identified sources of resistance to Southern rust in Latin American germplasm. Since these new germplasm complexes were merged with local germplasm, no further epidemics have occurred.

Considerable international assistance has been provided to West and Central African countries to build up national maize research programs. CIMMYT and IITA, at different times, have participated in major development assistance projects in Zaire, Cameroon, Ghana, Benin, and Burkina Faso to develop and strengthen national maize research programs. The international centers have stationed resident maize research staff in countries of West and Central Africa, helped to build experiment stations, and provided considerable training and scholarships for graduate student training. The largest publicly funded national maize research effort is in Nigeria. Ghana's research program has been especially vigorous since the 1980s (see Chapter 10).

North Africa and the Middle East. In Egypt and Turkey, national maize research programs were established by the ministries of agriculture during the 1950s, with assistance from the USDA, the U.S. foreign aid program, FAO, and the Rockefeller and Ford foundations. A number of double-cross hybrids were developed from American inbred lines and local and exotic germplasm sources. These new hybrids turned out to be highly susceptible to late wilt disease in Egypt, which devastated them. Even after a resistant double-cross hybrid was released in 1965, farmers remained reluctant to grow hybrids, preferring their improved open-pollinated varieties, which had greater resistance to late wilt. Today, maize research in Egypt is carried out by the national maize program of the Ministry of Agriculture, Cairo University, and several private local and transnational maize seed companies.

In Turkey, national maize scientists initially relied exclusively on U.S. hybrids and inbred lines. Later, they developed local materials and crossed them with elite U.S. maize germplasm. During the 1970s, the national maize program began to give more emphasis to open-pollinated varieties. After making headway, hybrid development work was renewed in 1980.

During the 1980s, many private seed companies with international operations have established maize seed production programs in Turkey based mainly on hybrids developed elsewhere. Over time, it is expected that these companies will also engage in maize breeding research within the country. With the growth of the private seed sector, the role of the national maize program is being reviewed. It appears that hybrid maize development will be left to the private sector, while the public-sector maize research program will concentrate on developing improved open-pollinated varieties for the more marginal production areas.

South and Southeast Asia. National maize research in South and Southeast Asia received a major impetus from the Rockefeller Foundation, which established a maize research program in 1957, in cooperation with the government of India.

National maize research programs in South Asia are primarily located at federal and provincial research institutes and, in India, at public universities. Since the 1980s, private maize improvement research (by local and transnational companies) has expanded rapidly, especially in India, in response to rising demand for hybrid seed to plant winter and spring maize.

In the Philippines, agricultural universities and colleges and national research institutes have had maize research programs since the 1950s. Maize research in the region got a major boost when the Rockefeller Foundation sent scientists to Southeast Asia in 1967 to help formulate research strategies to bring downy mildew under control. As a result of collaboration with scientists in Thailand and Philippines, varieties resistant to downy mildew were developed rapidly and widely diffused among farmers during the 1970s.

The national maize research systems in Thailand (see Chapter 10) and the Philippines are particularly strong. Research is undertaken by national institutes, public universities, and local and transnational seed companies. In Vietnam in recent years, the national maize research system has grown in size and impact.

East Asia. East Asia accounts for approximately 40 percent of Third World maize production. China and North Korea, which have most of the maize in this region, have well-developed national research systems, as shown by their high average yields and rates of production growth. The Chinese maize research system is discussed in Chapter 10. South Korea is now increasing resources devoted to maize research at universities and national institutes in hope of satisfying more of its demand through national production.

5

Research Policy Issues

Policy makers who set out to modernize their maize economies by choosing optimum research investments have to confront a wide range of issues and resolve many difficult questions. Fortunately the global maize research system formed in recent decades offers many sources of genetic materials and information. The stage of development of the maize program and economic and environmental conditions influence the choices planners must make on key issues such as targeting farmer groups, importing varieties, developing hybrids, and conducting basic research. A strategic plan can help the maize program focus its resources on the most critical research and development priorities. The goals identified in the strategic plan can best be achieved by revamping outdated organizational structures that enfeeble research activity and by ensuring that the research organization becomes involved in the technology transfer process.

Accessing the Global Maize Research System

An intricate matrix of research organizations—public and private, national and international—is engaged in maize research of relevance to Third World production conditions. These interconnected research organizations, especially ones in the public sector, form a global maize research system. For policy makers in developing countries, the system offers alternative sources of research information and germplasm and, therefore, a wide range of options for investing in maize research.

In developing countries, most maize research has been undertaken by public organizations. Today, more than 60 developing countries have national maize improvement programs carried out by government departments, crop research institutions, and agricultural universities. During the 1990s, public-sector research programs will remain important sources of new maize technologies in most developing countries. Particularly for marginal production environments characterized by stresses such as frequent drought or waterlogging, acidic soils high in free aluminum, and various soil nutrient deficiencies, it is likely that national programs, with backstopping from international agricultural research centers, will continue to make the principal research investments in new maize technologies.

Private seed companies, both transnational and national, are poised to become important suppliers of hybrid maize seed and other research products and information. Private seed companies will focus on the geographic areas and farmers that offer the greatest economic opportunity. Their research investments will tend to focus on areas that are likely to be profitable fairly quickly. For private companies to engage in much long-term research, seed sales must reach a substantial volume—and the profit potential must be adequate.

Where the ecological and economic conditions are favorable to the modernization of maize production, developing countries should enact policies that foster the development of private seed industries. Doing so will have implications for future research agendas of public maize research organizations and the international centers that serve them.

Historically, most private maize seed companies began by producing and distributing seed developed by publicly funded maize programs. As seed sales increased, some companies began to invest in proprietary maize research. Today, the pattern has changed somewhat. Some transnational companies like Pioneer, DeKalb, and Cargill are conducting maize research for tropical and subtropical environments, though they have yet to attain sizable volumes of seed production and sales. But even these large companies are making extensive use of public-sector germplasm from international and national maize research programs.

Many of the technological components that publicly funded national programs have developed will be useful to private seed companies. Therefore, public maize research organizations need to develop policies for sharing their research information and products with the private sector. The best policy is to make research information and products available to all private organizations on a uniform, nonexclusive basis. This is not to say that public-sector organizations cannot charge royalties for access to their research products; they certainly can. However, royalties should be set in such a way that reputable private firms have fair access to the research innovations of publicly funded institutions.

The minimum conditions necessary for the development of private maize seed enterprises are protection of private property, opportunity to earn satisfactory profits, freedom to operate without undue interference from government agencies, and protection from unfair competition such as seed sales by a subsidized parastatal company or government favoritism toward certain private companies.

Private firms, mainly from industrialized countries, also undertake research to develop new agricultural chemicals for control of weeds, diseases, or insects in maize, as well as to develop machinery and equipment to improve efficiency in land preparation and management, application of inputs, weed control, and harvesting. Cost-sharing partnerships between private companies and public research and extension organizations can accelerate the diffusion of these technological components. Such collaboration should be encouraged.

CIMMYT and IITA will remain the most important international sources of improved germplasm and research information for national maize programs in developing countries that have tropical and sub-tropical maize-growing environments. CIMMYT (globally) and IITA (regionally) are hubs of international maize germplasm exchange, information, and training networks that link thousands of maize researchers and their organizations. Over the nearly 30 years that these centers have been in operation, they have generated large stocks of maize research information and products.

Yet changes are under way that will affect how CIMMYT and IITA interact with national programs. First, CGIAR funding for maize research, in constant dollars, is declining. Second, both centers are reducing their activities in applied and adaptive research as strong national programs and private seed companies expand their activities. Third, both centers are moving their research agendas "upstream" That is, they are seeking to provide a wider range of tools and knowledge to improve the research impact and efficiency of national maize research programs, and they are de-emphasizing the development of more finished technologies and genetic materials.

CIMMYT and IITA view publicly funded national maize research programs as their primary clients. However, they also see the private sector as an important component of national maize research systems because private seed companies are alternative suppliers of improved germplasm. In meeting seed requests from the various research groups that make up national maize research systems, both centers give priority first to publicly funded national programs and then to private seed companies within client countries and transnational private seed companies. If requested by the government, CIMMYT and IITA will channel all their germplasm deliveries to a country through the publicly funded national program.

As a nation's private sector becomes capable of supplying good seed to a large portion of the maize growers and begins to conduct proprietary research, policy makers will have to reorient public maize research. One obvious change is to de-emphasize public-sector development of varieties and hybrids for the more-favored maize production environments. Of course, public institutions should continue to develop varieties for the more marginal production environments and for classes of farmers not likely to purchase hybrids. Public research organizations will also need to continue longer-term work on basic source germplasm to support and complement private maize breeding research. However, many entrepreneurial seed companies in developing countries that are too small to do all of their own research will continue to depend on national research systems and international centers for improved open-pollinated varieties, early generation lines, and, probably, advanced inbred lines. It is likely that these small companies will be a vital part of national maize seed industries.

When planners move to stimulate private maize research activity, they may encounter resistance from public research organizations that fear smaller budgets and limitations on the scope of their research. Naturally, it is not easy for public research institutions to give up traditional program responsibilities. And, it may be less satisfying to public-sector researchers to limit their work to development of basic source germplasm and the problems of maize production in marginal areas. However, if public research is not redirected as private research expands, unnecessary duplication and competition will dissipate scarce funds that could be better invested in other research.

Development Stages of National Maize Research Systems

The evolution of public and private research collaboration in the United States has considerable applicability for most Third World countries, especially since the end of the cold war and the discrediting of command economies in favor of market-oriented ones. The United States has a long history of fruitful collaboration between public and private organizations in maize research and technology development and diffusion. Private enterprise in the United States has been especially active in marketing of fertilizer and other agricultural chemicals, in hybrid seed production, and in farm equipment services.

Figure 5.1 schematically portrays the changing roles and responsibilities of public and private breeding and seed production organizations at different stages of development in a national maize economy. Obviously, this model does not necessarily fit the national development path of any individual country. Still, we believe that the desirable and cost-effective development path for most countries begins with public research and production organizations that initiate the process of development in the maize economy and then gradually transfer many activities to private organizations as the maize economy advances.

Stage		Basic germplasm development	Final product development		Production of foundation seeds		Production of seeds for farmers	
			OPVs	Hybrids	OPVs	Hybrids	OPVs	Hybrids
1: Most maize producers are subsistence farmers who use few purchased inputs. Maize yields are low (less than 2 t/ha).	Public national	■	■■■		■■■		■■■	
	Private sector							
	IARCs	■■■	■■					
2: Most maize producers are subsistence farmers, but increasing numbers have higher level of management. Maize yields average 2 to 3 t/ha.	Public national	■	■■■	■■	■■	■	■■■	■■
	Private sector			■■■		■■■		■■■
	IARCs	■■■	■					
3: Half the maize producers are commercial; maize is a major source of income. An adequate infrastructure for input-supply and commodity markets exists. Yields average 3 to 4 t/ha.	Public national	■■	■■■	■	■■	■	■■	
	Private sector			■■■	■■	■■■	■■■	■■■
	IARCs	■■■						
4: Most producers are commercial. Infrastructure for input- supply and commodity markets is excellent. Yields frequently exceed 4 t/ha.	Public national	■■■						
	Private sector	■■		■■■		■■■		■■■
	IARCs	■■■						

FIGURE 5.1 Schematic stages of development of national maize programs: Responsibilities of public national programs, the private sector, and international agricultural research centers (IARCs). ■ = small activity. ■■ = significant activity. ■■■ = major leadership.

In stage 1, virtually all maize research and development activities are carried out by publicly funded national programs and international agricultural centers. Average maize yields typically are low (less than 1.5 t/ha), and most maize farmers are small-scale subsistence producers who use few purchased inputs. The final product of maize breeding research is mainly improved open-pollinated varieties, the seed of which can be produced by either public seed organizations or small private growers. Improved varieties supplied by the international centers have a major role in final product development. Public national programs focus on selecting superior experimental varieties from materials of the international centers for release to farmers. Numerous developing countries, especially in Africa but also in poorer parts of Asia and Latin America, are in stage 1 and the transition to stage 2.

In stage 2, the public sector remains dominant in basic germplasm development, although final product development has expanded beyond open-pollinated varieties to hybrids. Farmers, whose average yields typically exceed 1.7 t/ha, have begun to adopt productivity-enhancing technologies. In this stage, the relationship between the international centers and national maize research organizations has changed. The international centers now tend to contribute earlier-generation breeding materials to the national maize program for final product development. The national program produces both open-pollinated varieties and hybrids, and public organizations and small private growers are still primarily responsible for foundation and commercial seed production. However, the private sector also is engaged in hybrid seed production, mainly using public-sector inbred lines and hybrids. Countries such as Thailand, Philippines, Mexico, Peru, Tanzania, and Nigeria typify stage 2 of development.

In stage 3, international centers are no longer supplying final products to national maize research institutions. Rather they contribute new germplasm complexes to support the final product development work of national programs and private companies. Public institutions are still engaged in developing open-pollinated varieties for marginal maize-growing areas along with developing some inbred lines and hybrids. Publicly supported seed activities are confined to the multiplication of open-pollinated varieties and foundation seed production of public inbred lines. In this stage, private companies dominate final development of hybrids as well as commercial hybrid seed production. Countries such as Egypt, Turkey, and Brazil typify stage 3 of development.

In stage 4, public-sector programs no longer engage in final development of maize varieties or hybrids. Instead, along with the international centers, they only develop basic germplasm complexes (new source populations to supply specific traits and to increase genetic variability) in support of private maize research organizations. The final development of hybrids, as well as foundation and commercial seed production, is now completely in private hands. Countries such as Argentina, Chile, and Zimbabwe typify stage 4 of development.

Two important cautions need to be given about this simplified maize research development model. It implies a uniformity in the environments and circumstances of maize farmers that seldom exists. In most developing countries, the maize economies are, at the least, dualistic— made up of both subsistence and commercial sectors. Thus, where the national maize area is large and the potential exists to increase the productivity in substantial portions of this area, the conditions may be right for the private sector to assume increasing responsibility for serving the more-favored production areas and farmers, while public-sector organizations continue to be the primary source of improved maize research information and products for the less-favored production areas and farmers. Thus, it is likely that within a particular country the national maize research system will operate in several stages simultaneously, depending upon the region, ecology, and target farmer groups.

The second caution is that the private sector can be divided into two groups—large companies, often transnational, which tend to serve the most-favored hybrid production areas, and small companies, usually locally owned, which tend to serve the less-favored production environments where competition from large companies is less intense. Over time, the larger companies will increasingly undertake proprietary maize breeding research. The smaller companies will continue to rely on public maize research programs and the international centers. Even in the United States, many small seed companies, and some of the larger ones, depended on public-sector germplasm for development of varieties and hybrids until the late 1970s. Now, private foundation seed companies have largely replaced universities as suppliers of inbred lines, although publicly funded maize research programs continue to contribute elite germplasm widely used by maize breeders in private companies.

Issues in Maize Research Planning

In addition to the larger institutional issues related to public and private maize research investments, organizational planners must consider a broad range of issues in establishing their research strategies and operational plans.

Strategic Planning

Strategic planning in research and development institutions has been widely written about elsewhere, but a few guidelines can be given here. Strategic planning describes the institution's vision of the future and the operational principles that will get it there. Institutional planning should be more than an extrapolation of previous strategies and activities. It should take a fresh look what most needs to be done and how. In addition, it is important to distinguish between strategic planning and operational planning. By nature, strategic planning is a top-down activity while operational planning is a bottom-up process. Despite these differences, a strategic plan will not succeed unless those responsible for carrying it out become true stakeholders. Operational planning defines how the strategic objectives will be pursued within a given period.

A strategic planning exercise begins with an assessment of current strengths and weaknesses of the maize research organization. The organization has to decide what it will seek to do and what it will leave to others to do. Human and financial resources, obviously, will weigh heavily in such considerations. It is important to define clearly the expected products of the maize research program and the intended clients and end users of these products. Often the client of a maize research program is the extension service or some other organization responsible for technology diffusion. Although the end user, inevitably, is the farmer, the term may be too general. A particular category of farmers, e.g., resource-poor farmers or women farmers, may be the target end user for some maize research organizations. Sometimes the product is not a new technology, but rather a new research methodology, and the clients are other maize scientists.

In establishing the maize research priorities, the capacity of science to overcome problems constraining increased maize productivity should weigh heavily. Maize research plans should be based upon well-defined objectives and have a clear product orientation—the development of specific products for dissemination to target groups of farmers. This demand-driven planning approach is typical of private maize research enterprises but is rarely employed in publicly funded institutions where research plans tend to center on topics such as germplasm improvement, agronomy, pathology, and entomology.

A maize research program needs to develop a mission statement that summarizes the work it will do and describes, in general terms, how its program will function. This statement should be short, covering only the main program (business) areas and the guiding philosophy and operational principles. Based on the mission statement, a set of strategic planning criteria and operational objectives need to be established for making decisions about future operational strategies. From these criteria, a series of operational research projects and activities can be planned that are consistent with the mission statement. Operational plans, unlike the strategic plan, must take into account realistic estimates of what can be accomplished within fixed time periods and within funding and human resource constraints.

Production and Environmental Considerations

One important issue in planning research investments is the priority that is given to raising productivity in the better maize-growing environments versus the more marginal ones. Obviously, the better environments, especially if they have been underutilized, offer the greatest near-term potential for raising maize productivity and production. There are powerful economic and environmental reasons for increasing the intensity of maize production in ecological conditions that lend themselves to intensification, while decreasing the intensity of maize production in unfavorable ecological conditions.

But because of population growth in most developing countries, maize cultivation has spilled into marginal lands that are adversely affected by moisture stress, extreme temperatures, low solar radiation, various diseases, insects, and weeds, and soil infertility and toxicity problems. Therefore, from a social welfare standpoint, the economic betterment of resource-poor farmers in these marginal production areas may be of considerable importance to a national government. However, T. W. Schultz, the Nobel Laureate agricultural economist, observed that research and investment priorities made solely on the basis of equity considerations usually end up retarding the rate of improvement in agricultural productivity. Often such maize research schemes backfire because they go against the laws of comparative advantage in trying to reach disadvantaged rural groups.

Notwithstanding Schultz's warning, the historical spread of maize into nontraditional environments—made possible by its broad adaptation and through advances in science—suggests that researchers should continue to try to develop improved germplasm and production technology that will permit maize to be produced in environments that may previously have been considered unfavorable for economic production. However, planners should recognize that some ecological stresses in marginal agricultural areas are too severe for science to overcome economically and thus they do not merit much research work.

Costs must be carefully weighed in determining how many production environments receive research attention because each ecological zone will have unique production problems and potentials. Normally, this means that a maize experiment station will be needed—with the requisite field and laboratory facilities—for each distinct environment targeted for research.

Targeting Distinct Farmer Groups

The vast majority of maize producers in developing countries are small-scale farmers, many of whom also can be described as resource-poor farmers. But the needs of both groups and the potential technological solutions are not identical. A distinction must be made between the resource-poor and small-scale producers because the latter may have good quality land and water resources but only in small quantities. In serving the small-scale farmer, higher labor productivity and income should be emphasized. Basically this calls for the development and introduction of yield-increasing, cost-reducing technologies. High-yielding, management-responsive open-pollinated varieties and hybrids, fertilizer use, and more intensive cropping sequences are examples of improved technologies that are appropriate for adoption by small-scale producers. But hybrids may not be suitable for resource-poor maize farmers, especially those who face severe environmental stresses or unpredictable markets and prices for their surplus harvest.

Even for resource-poor farmers in marginal environments, a key investment criterion should be the potential for creating new technologies that have much higher yield potential and profitability than the current production practices. Unless researchers can create substantial and economically exploitable yield gaps in marginal areas, there is little prospect that farmers can be induced to alter their production practices.

Importing vs. Breeding Domestic Varieties

In large maize-producing countries—those that have more than 100,000 hectares of maize—a national maize breeding program is usually economically justified. Considerable benefit can nevertheless come from policies that allow the testing and importation of adapted varieties developed outside the country.

CIMMYT and IITA are major sources of improved germplasm. These international centers historically have given priority in their germplasm distribution (especially advanced generation, more refined materials) to public-sector maize research organizations. However, they also have supplied portions of their breeding materials to private maize breeding firms. Some large public research programs (United States, France, Yugoslavia) also supply their improved maize germplasm free of charge to national breeding programs in developing countries.

In countries where human and financial resources for maize research are very limited, the emphasis should be on testing germplasm from other organizations (usually varieties and lines developed by national and international maize breeding organizations). Even for advanced maize research programs, germplasm developed elsewhere will be an invaluable source of the basic building blocks for improved varieties.

Plant quarantine regulations regarding the flow of breeding materials to and from the country should be as simple as possible, while still providing protection against the introduction of plant diseases, insects, and weeds not already present in the country. Unnecessary delays and tests retard research progress and slow the development of a dynamic and profitable maize seed sector.

To encourage private maize breeding, the national government must establish policies that facilitate the importation of breeding stocks and seed for multiplication. Private seed companies should not be expected to reveal detailed information about the pedigrees of materials they import or develop. Without such protection, they have no incentive to invest in maize research.

Access to Public Lines and Varieties

In the past, access to publicly developed genetic materials was not an issue because national research organizations, agricultural universities, and the international agricultural research centers were producing most of the improved maize germplasm in the Third World. But private seed companies are now springing up in many countries. Consequently, national governments need to establish even-handed, nonexclusive policies to promote diffusion of improved genotypes among farmers. Public national and international maize breeding programs should make improved varieties, inbred lines, and hybrids they have developed available to private seed companies. And they need to have adequate facilities and staff for multiplying breeder and foundation seed to supply to private seed enterprises.

At the same time, public research organizations can reasonably expect to directly recover some of their investment from companies that market seed based on their germplasm. The companies should pay a royalty or user fee to the public research organization for seed production of their finished products but not necessarily for use of early generation improved germplasm in breeding programs. The fees should be structured to promote wide adoption of public-sector improved seed by farmers and the development of a profitable maize seed industry involving many seed producers.

Hybrids vs. Open-pollinated Varieties

There is a long-running research debate about whether open-pollinated varieties or hybrids are best for small-scale farmers. The debate centers on the greater yield and dependability of hybrids and the prospects for developing hybrid seed industries to supply small farmers.

The yield superiority of hybrids is clear in the temperate zones. Compared with improved varieties from the same genetic background, single-cross hybrids typically yield 30 percent more and double-cross hybrids yield 20 percent more. Such yield advantages are not yet evident in most tropical environments. In trials in lowland areas of Central America, for example, single-cross hybrids only had about 15 percent yield advantage over the best improved open-pollinated varieties developed through rigorous and effective forms of recurrent selection.

Hybrid maize, mainly from the United States, gained a spotty reputation in tropical countries during the 1950s. The hybrids that were introduced were not well adapted to hot environments, lacked appropriate disease resistance, and consequently yielded poorly. In addition, hybrid seed production was entrusted to public-sector organizations, most of which proved incapable of producing and distributing quality seed to significant numbers of farmers.

As a result, maize researchers in most developing countries concentrated on developing better methods of selecting high-yielding open-pollinated varieties with good uniformity and yield dependability. These improved varieties often had much better yields than local landrace varieties. Although the yield potential was 10 to 30 percent less than that of the counterpart hybrid, the fact that the farmer could save the seed of the open-pollinated variety for replanting the next year, with little loss in yield potential, made the task of seed production and distribution much easier.

The success of improved open-pollinated varieties discouraged developing countries, especially in the tropics, from investing significantly in hybrid research until the 1980s. IITA and CIMMYT now have developed a broad range of superior germplasm for use in developing high-yielding open-pollinated varieties as well as inbred lines for hybrids. Private seed companies are using this germplasm to develop hybrids for tropical conditions, and that may result in the release of superior hybrids adapted to tropical conditions. Proponents expect tropical and subtropical hybrids to possess the same 20 to 30 percent yield advantage over open-pollinated varieties (even in stressed environments) that is evident in temperate climates of industrialized countries.

Another aspect of the argument about hybrids and open-pollinated varieties is the belief that by promoting open-pollinated varieties, maize researchers have unwittingly retarded the establishment of viable maize seed industries, which are a prerequisite for modern maize economies. One reason is the difficulty a seed company has in forecasting demand for seed of open-pollinated varieties. Farmers do not have to purchase seed of open-pollinated varieties every season because they have the option of replanting the last season's seed, usually without suffering much yield depression. Although some countries, such as Thailand and Egypt, have developed successful maize seed industries that produce primarily open-pollinated varieties, they may be special cases. In both countries, open-pollinated varieties were the only high-yielding genotypes available that had resistance to economically devastating diseases that were present (downy mildew in Thailand and late wilt in Egypt).

Eventually hybrids may be the preferred varietal types for commercial sale throughout the developing world. But we believe that a dynamic and successful maize seed industry can be built by the public and private sectors through offering a progression of commercial genotypes, beginning with improved open-pollinated varieties, followed by higher yielding nonconventional hybrids, and eventually emphasizing conventional double-cross, three-way cross, and single-cross hybrids.

High-input vs. Low-input Maize Technologies

Some maize researchers contend that producing technologies that require few purchased inputs and emphasize yield stability over yield potential is the most appropriate approach for serving small-scale maize producers who face great risks, especially in marginal production environments. These are valid research objectives particularly for environments that have frequent droughts or widespread soil infertility problems. Nonetheless, even for these environments, it is possible to develop hybrids that are tolerant of low-input management, but have the potential to yield well under high-input management.

Applied vs. Basic Research

Another contentious issue is to what degree maize research institutions in developing countries should invest in basic research, such as biotechnology using molecular probes and other genetic engineering tools. Our view is that because of its costliness, high standards of precision, long-term nature and high risk, biotechnology research will be increasingly concentrated in industrialized countries. Moreover, although considerable amounts of basic background molecular research will be undertaken by publicly funded organizations, it is likely much of the more applied biotechnology research will be undertaken in the private sector. Access to these innovations will be determined by market forces.

Thus most developing countries will be importers and modifiers, rather than generators, of advanced technology in the coming decades. This should not worry research planners in the developing world. Substantial research gains can be made through conventional plant breeding techniques. In addition, until maize technology delivery systems become more effective, widespread adoption of advanced-technology maize products will not be practical. Finally, there is no reason to believe that developing countries will be denied the results of maize biotechnology research, providing they can participate in the marketplace and have the requisite infrastructure to use it.

Organization of Maize Research

Interdisciplinary Maize Research

Maize research requires various types of scientific expertise organized in a way that integrates the researchers. In most instances, maize research is best done by commodity-focused interdisciplinary teams. Interdisciplinary maize research programs should include scientists from such disciplines as breeding and genetics, plant pathology, entomology, physiology, agronomy, and economics. Research teams are able to focus on the multiple facets of maize research and development problems, unlike individual researchers who tend to see a problem in isolation, that is, from the vantage point of his or her own specialty.

Unfortunately the most common research structure in developing countries places the scientists in disciplinary departments where they work on maize from their disciplinary perspective, but participate in some form of a commodity-specific maize research program. The chief weakness of such a structure is that the scientists end up being responsible to two very different bosses: their disciplinary department head and the maize research program team leader or coordinator. Often the scientist finds that important personnel matters like performance evaluations, salary increases, and budget control are exercised by the department head, but the program of work and day-to-day supervision is set by the maize research program leader. Another weakness is that although individual disciplines may be adding to the stock of scientific knowledge, the information being generated is not being integrated into new technologies that overcome farmers' yield constraints.

Maize Breeding vs. Crop Management Research

Although the development of high-yielding germplasm has been the catalyst for modernization of maize production, the amount of research funds that has flowed to maize breeding research is out of proportion to the funds allocated to crop management research. As a result, the research products (germplasm) of most national maize improvement programs are at a more advanced stage than the research information produced by the crop management (agronomy, economics) programs. One explanation is that an effective national maize breeding program can be mounted and managed with fewer scientists than a dynamic crop management research program involving extensive on-farm research. Second, because plant breeding is mainly carried out on experiment stations, it is less costly and easier to supervise than a crop management research program, much of which must be conducted off the station in farmers' fields. Third, because the international centers are primarily plant breeding institutions, national maize breeding programs have found more support and backstopping for breeding programs than for crop management research. Fourth, researchers have difficulty defining the product needs in agronomy and economics research because crop management research has more facets than plant breeding.

Still, there is little doubt that improved crop management will be the principal sources of higher maize productivity in farmers' fields over the next several decades in the developing world. And, the need to sustain the productivity of natural resources will add to the pressure for increased investments in crop management research.

On-station vs. On-farm Research

Too often research information and products are inappropriate to the circumstances of typical farmers. A major reason is that most national maize research programs conduct too much research under the controlled conditions of research stations and not enough under the more-realistic production conditions of farmers' fields. As a first step in expanding maize research on farmers fields, the maize production problems facing farmers should be identified and ranked in importance. This information should then be used to establish priorities for both crop improvement and crop management research. Finally, possible new maize varieties and crop management technologies should be subjected to rigorous on-farm testing and evaluation before being officially recommended for extension to farmers.

The Need for Economics Research

In most national maize programs, practical, production-oriented economics research is either nonexistent or drastically underfunded. Economists should be integral parts of interdisciplinary maize research teams. They should routinely assess the economic circumstances of target farmer households as well as the profitability of recommended maize production practices. They should know the physical response of recommended maize varieties to modern inputs under diverse conditions; what input and output prices farmers face and what their effects on economic profitability are; whether the lack of knowledge of those inputs is the real barrier to farmer adoption or whether the constraint is lack of supply of the right inputs at the right prices at the right places; whether farmers have their own savings or whether institutional financing is needed; and whether the market, transport, storage, and processing infrastructure retard development of the maize economy.

Linking Maize Research and Technology Transfer

Few national agricultural research systems in the developing countries have adequate program linkages to the production organizations that engage in technology transfer. Technology transfer is generally assigned to agricultural extension and input-supply organizations. Some research planners argue that maize research institutions should have only limited responsibility in technology transfer on the grounds that other organizations are better suited to do such work and that researchers should concentrate on making maximum advances in science. This argument has merit. The problem is that in many developing countries the research system generates improved technology that rarely reaches farmers' fields because the technology transfer organizations are ineffective. Under these imperfect institutional circumstances, maize research organizations must participate in technology transfer.

There are many ways in which a maize research institution can help to strengthen the technology delivery system. One is to produce superior new maize technologies. Another is to provide strong training and technical backstopping to the extension service and other groups that are attempting to demonstrate and introduce the new technology. (Where the private companies are involved in input delivery, they should share in the costs of maize technology demonstrations with public research and extension organizations.) Also, maize research organizations should conduct economic policy research and public-information programs aimed at government planners and political leaders.

The development of superior technology is not in itself sufficient to ensure adoption by farmers. New technology must be wedded to appropriate economic policies in order to maximize the return to maize research investments and to modernize the maize economy rapidly. In most developing countries, unfortunately, the links between agricultural research organizations and those responsible for formulating economic policies are weak at best. Therefore, when maize researchers have developed superior maize technology, they need to make their case forcefully to policy makers to bring the results of the research to fruition.

To be in a position to influence agricultural policy, maize research programs should generate and publicize information on the suitability and economic potential of various technological components for specific environments and groups of farmers. Also, maize researchers should be involved in monitoring on-farm technology demonstrations undertaken by production organizations and in evaluating outcomes. And, maize research programs should regularly issue reports to publicize the adoption and impact of improved maize technology and to call attention to key bottlenecks that impede widespread and rapid dissemination such as input delivery, marketing channels, and prices.

Tanzania. Photo courtesy of CIMMYT.

6

Maize Genetic Resources

Importance of Genetic Diversity

Genetic variability is the lifeblood of any species and the essential raw material used by the plant breeder in crop improvement. The saying, "variation leads, breeders follow," has much truth to it. Maize exhibits great genetic diversity in plant, ear, and seed characteristics, in resistance to diseases and insect pests, and in tolerance to various environmental stresses.

The wild ancestors of maize evolved through natural selection caused by environmental stresses and by natural mutations and recombinations until its domestication by man 6,000 to 8,000 years ago. In the Neolithic period, farmers converted the plant from a hardy perennial wild type with relatively low grain yield and an unprotected ear into a high-yielding annual plant with a large, fully sheathed ear unable to distribute its seed, and thus survive, without the aid of humans.

Since domestication, maize has evolved through parallel processes of natural and farmer-breeder selection. The maize plant has adapted to many biotic and abiotic stresses, such as cold, heat, drought, certain soil toxicities, diseases, and insects, enabling it to survive and thrive in a diverse range of environments. From its center of domestication in Mexico, maize spread throughout the Americas and, beginning in the 16th century, through Europe, Africa, and Asia.

With the rapid diffusion of maize hybrids in the United States during the late 1930s and 1940s, older open-pollinated varieties and indigenous strains largely disappeared from American farms. And when maize breeding programs were launched after World War II in Latin America and Asia, it seemed likely that hybrids would also replace many native strains and varieties. Concern that the loss of those genetic resources could limit future maize improvement and deprive plant breeders of genes of great potential importance led to organized efforts to collect, preserve, and classify native maize varieties and strains—the products of thousands of years of evolution.

In 1943, scientists in the Mexican-Rockefeller Foundation cooperative agricultural program began collecting Mexican maize landraces to identify outstanding cultivars and germplasm complexes for use in breeding programs. Although collecting and preserving genetic diversity was not the original objective, its value was soon recognized. Aided by Rockefeller Foundation programs in Mexico, Colombia, and Central America, germplasm collection and classification work quickly expanded into an international campaign led by several national maize programs to collect maize varieties throughout Latin America and the Caribbean.

In 1954, the U.S. National Academy of Science (NAS) formed a special committee to promote the collection and preservation of indigenous strains of maize in the Americas. With its support, over 10,000 collections were identified and more than 200 races were described. To store seed samples from most of these collections, germplasm banks were set up in Mexico, Colombia, Peru, and Brazil. Small duplicate samples from all NAS-sponsored germplasm collection activities also have been sent to the U.S. National Seed Storage Laboratory at Fort Collins, Colorado, for long-term storage and safekeeping. During this same time, national maize scientists in Italy, Romania, Spain, and Yugoslavia and Japanese maize scientists in Nepal were active in collecting and classifying local maize germplasm, which added considerably to the knowledge of the evolution and migration of maize in those areas of the world.

Today, 300 races of maize, consisting of thousands of different cultivars, have been identified and described around the world. Maize researchers estimate that these collections represent 90 to 95 percent of the genetic diversity of the maize species. Most of these races are tropical in adaptation. Paterniani and Goodman (1977) reported that 50 percent of the races are adapted to low elevations (0 to 1,000 m) and 40 percent to high elevations (above 2,000 m), with the remainder adapted to intermediate elevations. They also reported that about 40 percent of the races have floury endosperms, 30 percent are flints, and 20 percent are dents. The remainder are popcorns and sweet corns.

The network of germplasm banks that holds seed samples of most of the primitive cultivars and old landraces and varieties that have been collected in the past 50 years is an important source of maize genetic diversity. Another source of genetic diversity is the open-pollinated varieties that farmers in age-old centers of cultivation still grow. Despite the hybrid revolution, unimproved and improved open-pollinated varieties still occupy nearly 60 percent of the maize area in developing countries. A third source of genetic diversity, and one which many breeders argue is the source of the most useful germplasm for conventional maize improvement, is maize breeding programs around the world. The genetic resources of these programs include inbred lines, hybrids, open-pollinated varieties and synthetics, gene pools and populations, interspecific crosses, and a diversity of cytoplasmic sources. Although containing useful genetic diversity, these breeding materials represent only a small fraction of total maize germplasm pool (less than 10% of all races), indicating that much genetic diversity in the genome remains to be exploited. The expanding work in molecular markers, gene transfers, transformations, and other biotechnology research may lead to the creation of much new genetic diversity that could serve as source materials for future improvement of specific maize characteristics.

Obtaining Maize Genetic Resources

Publicly funded and privately funded maize research operations around the world are collectively improving a vast array of maize germplasm complexes. Most of the global maize research is publicly funded and the products and information generated are widely accessible to scientists worldwide. But during the past 30 years, driven by the hybrid revolution, private investments in maize research, especially plant breeding, have steadily increased.

Individual scientists and maize research organizations can obtain samples of maize genetic resources from other national programs and from international programs and germplasm banks. A central function of the international agricultural research centers has been to serve as hubs for the preservation and utilization of genetic resources, for germplasm exchange, and for the operation of testing networks for the major food crops including maize. The germplasm testing networks have brought together the work of thousands of scientists and hundreds of organizations. They have accelerated progress in plant breeding through the worldwide distribution of the best germplasm generated by the publicly funded international and national plant breeding programs.

The U.S. Department of Agriculture (USDA) and several other national agricultural research systems from such industrialized countries as Yugoslavia, France, Italy, and Japan have also followed policies of relatively widespread and free distribution of improved germplasm to maize scientists in other countries. Breeding programs in developing countries have made much use of the improved germplasm developed by U.S. maize breeders at state agricultural universities and USDA research facilities and by European maize breeders from publicly funded research programs.

Access to private research information and intermediate products is much more restricted. To assure a good return to investment, private maize seed companies view their research products and information as proprietary. For other breeders, access to a private company's superior germplasm, therefore, comes only through contractual acquisition of proprietary breeding materials or in the purchase of commercial hybrids that the company produces and distributes.

Maize Germplasm Banks

Germplasm banks grew out of the need to store and preserve the extensive maize genetic resources that were collected and classified during the 1940s and 1950s. Many national maize research programs were unable to afford large germplasm bank facilities. Thus, it was evident that the seed viability of many existing collections would be threatened by poor storage conditions and faulty seed regeneration procedures.

Conservation of genetic resources is one of the mandates of the CGIAR system of international agricultural research centers. CIMMYT's maize germplasm bank began operation in 1970 and was stocked with duplicate seed samples from most of the maize racial groups collected under the Rockefeller and NAS programs conducted in Latin America. It also contained the best breeding materials (populations, varieties, gene pools) that had been developed by national maize research programs between 1940 and 1960 in Latin America, Asia, and, to a lesser extent, Africa.

The International Plant Genetic Resources Institute (IPGRI) has a mandate to (1) identify needs in exploration and carrying out field collection activities, (2) improve and safeguard storage of collected samples, (3) improve data storage and retrieval systems relating to genetic conservation and utilization, and (4) disseminate scientific information on plant genetic resources. IPGRI has supported systematic collection of maize landraces in more than 30 countries. Besides the seed held in national germplasm banks, duplicate seed from these collections has been stored at the U.S. National Seed Storage Laboratory or CIMMYT. IPGRI has published catalogs on national maize genetic resources (including evaluation data) for Argentina, Bolivia, Brazil, Chile, Paraguay, Peru, and Uruguay. IPGRI continues to support maize germplasm-collecting activities in areas of the world that are thought to represent gaps in the diversity of maize and its wild relatives.

Worldwide, 25 germplasm banks today contain 95 percent of maize's vast genetic diversity (Table 6.1). The seeds preserved in the global network of maize germplasm banks represent a diversity of types unequaled in any other cultivated species. The banks vary in size, in the geographic emphasis of their collections, and in their ability to preserve and regenerate seed and to document and evaluate germplasm collec-

TABLE 6.1 Important maize germplasm banks.

Institution	Accessions	Collection emphasis
VIR, Russia[a]	15,084	Europe
IMR, Yugoslavia	15,000	Europe
CIMMYT, Mexico[a]	11,100	Americas
INIFAP, Mexico	10,000	Americas
NSSL, USA	7,619	World
UNA, Peru	7,145	Americas
ICA, Colombia	5,000	Americas
RICTP, Romania	3,200	Eastern Europe
INTA, Argentina	3,444	South America
ISU, USA	3,000	North America
PGRC, Canada	2,800	North America
NIAS, Japan[a]	2,654	Asia
CIFEP, Bolivia	2,220	South America
CENARGEN, Brazil	2,500	South America
IPB, Philippines	1,678	South America
NBPGR, India	1,571	Asia
NARS, Kenya	1,500	Asia
CRIFC, Indonesia	1,368	Asia
MRI, Slovakia	1,306	Eastern Europe
INIA, Spain	1,040	Iberian Peninsula
CNU, South Korea	1,000	Asia
PGB, Portugal[a]	1,000	Iberian Peninsula
INIA, Chile	914	South America
TISTR, Thailand[a]	n.a.	Asia

[a]IPGRI-designated world or regional base maize seed collections.
Source: Plucknett et al. 1987; CIMMYT 1988.

tions. The Vavilov Institute in Leningrad, Russia, and the Maize Research Institute in Belgrade, Yugoslavia, have the largest collections. These banks largely contain germplasm collected in Russia and Europe. However, most of their accessions are inbred lines that contain much less genetic diversity than is found in the germplasm banks located in the Western Hemisphere, which have germplasm from Latin America. The most genetically rich germplasm collections are those maintained in Mexico by CIMMYT and INIFAP, at the National Agricultural University in Peru, at the National Agricultural Research Institute in Colombia, at EMBRAPA in Brazil, at INTA in Argentina, and at INIA in Chile.

CIMMYT has the best-organized and easiest to access maize germplasm bank and will supply small quantities of seed to maize research organizations without charge (though private companies are expected to pay shipping costs) in any country in the world. CIMMYT normally sends 50 seeds, although as many as 200 seeds can be supplied for more-intensive research. CIMMYT expects the organization receiving the seed to regenerate enough seed of each accession to meet its future needs. CIMMYT also asks recipients to send information on the performance of the accessions for inclusion in its global database.

During the 1980s, CIMMYT considerably strengthened its maize germplasm bank. The bank, headed by a full-time maize breeder, conducts research to classify and evaluate the CIMMYT collections with the hope of identifying useful germplasm for introgression into its conventional breeding programs. An expanded program of seed maintenance and regeneration has helped ensure that seed with good germination is available for all accessions. Seed regeneration programs have been established in Mexico at CIMMYT experiment stations located in different environments and outside the country through cooperative arrangements with several Latin American maize research programs.

CIMMYT also has made considerable investments in computerized database management systems to improve accessibility to the collections contained in the bank. A catalog containing "passport data" describing its bank accessions has been developed (available on CD-ROM disk), which is continually revised as new information about the accessions is generated. Scientists wishing to get seed samples from the CIMMYT germplasm bank should try to consult the catalog beforehand to improve the precision of their seed request.

The genetic resources of many national maize germplasm banks are difficult to access, despite their policy of supplying a reasonable number of seed samples to scientists on an apolitical basis. Seed distribution from national germplasm banks may be restricted because of inadequate budgets for seed regeneration and supply. Also, political constraints sometimes restrict scientific exchanges between counties. However, thanks in considerable measure to the efforts of IPGRI and other international centers, investments in genetic resource conservation have increased substantially in recent years.

In the past, many maize breeders have been skeptical about the value of maize germplasm banks. Some scientists consider them seed morgues of little practical value for breeding because more elite breeding materials often offer more useful and accessible genetic diversity. Even though primitive landraces are major sources of unique genetic variation for specific traits, tapping that resource is generally a time-consuming and uncertain activity. Breeders are seldom able to directly utilize obsolete cultivars and racial complexes as source material in breeding programs. Instead they first have to go through several cycles of enhancement and selection for adaptation and agronomic characters.

Another problem has been insufficient documentation of the adaptation and salient traits of each collection. Classification work is laborious and expensive, and it generates masses of data that are difficult to manage. In future as scientists learn more about the genome and as techniques to access useful gene combinations improve, more of the genetic diversity found in maize is likely to be utilized.

The low seed availability and viability of many germplasm bank collections due to weak seed regeneration programs is another barrier to their use by breeders. For the larger banks, operating an adequate program of seed regeneration is a formidable logistical task. Furthermore, many banks contain germplasm that needs to be regenerated in maize-growing environments not found in the country where the bank is located. International cooperation, therefore, is often necessary to maintain sufficient seed for use by plant breeders. The organization of such cooperation is often difficult for a national maize germplasm bank to undertake on its own, thus the need for international action to support seed regeneration.

International Maize Breeding Programs

Several publicly funded international and regional maize breeding programs operate in the Third World. CIMMYT and IITA maize scientists have developed broadly adapted gene pools and populations for tropical, subtropical, and highland environments from which improved open-pollinated varieties, inbred lines, and hybrids have been extracted. Both institutions focus on yield dependability. New germplasm source materials are being developed with enhanced tolerance or resistance to specific abiotic and biotic stresses. Both centers are producing early generation inbred lines adapted to the environments found in developing countries and they are generating information about heterotic patterns.

CIMMYT and IITA (in West and Central Africa) operate germplasm testing networks, sending thousands of packets of maize seed to more than 50 developing countries annually. As part of a code of ethics in international germplasm exchange and sharing, CIMMYT and IITA ask that collaborating scientists and institutions acknowledge the original source of breeding material supplied.

The CGIAR recently updated its policies on intellectual property rights and the associated plant breeder's rights and plant variety protection. Ways to maximize the unrestricted exchange and utilization of genetic resources and research production and information remain the guiding principle for these policies. The CGIAR centers act as a trustee for germplasm collections. The centers have an obligation to manage these resources for the benefit of humanity and to ensure access to them, without consideration of politics or financial gain.

The CGIAR's policies on intellectual property rights distinguish between natural plant genetic resources (landraces) and the germplasm products resulting from the centers' research. The output of the CGIAR centers' research programs is freely distributed and, wherever necessary, transfer agreements that protect the interests of developing countries are made. The CGIAR's preferred policy is the published announcement of genetic products to keep them in the public domain. Defensive and protective patents are sought only when necessary to ensure that developing countries have access to beneficial technology and products. Centers such as CIMMYT and IITA cannot seek financial returns from intellectual property rights conventions. If any financial returns arise from the application of intellectual property rights, these revenues are placed in an international fund for the conservation of plant and animal genetic resources.

National Maize Breeding Programs

More than 50 developing country governments operate maize breeding programs that have assembled germplasm source materials using local landraces and cultivars, introductions from other countries and research programs, and special source materials. Government research organizations differ in their policies on sharing breeding materials with other organizations. Most share seed with the international agricultural research centers free of charge. Further, assuming good inter-governmental relations, most national maize programs liberally share their germplasm with counterpart organizations in other countries.

The policies of government maize research organizations toward private maize seed companies, whether national or transnational, are much more variable and until recently have tended to be restrictive. Notable exceptions are the public maize research organizations in Brazil, El Salvador, Guatemala, Thailand, Turkey, and recently India, which have provided private companies ready access to breeding materials.

The trend today is for easier access by private seed companies to publicly developed germplasm and to breeder and basic seed for certified hybrid seed production. Kasetsart University's maize improvement program in Thailand is an example of a national program that has produced outstanding germplasm products and has been liberal in making its germplasm available to other public and private research organizations worldwide. For example, the population Suwan-1 and its inbred-line derivatives have been used extensively by public and private maize breeding programs in Asia for some time and, more recently, in Latin America and Africa.

Private-sector Maize Breeding Programs

Private maize research has steadily increased during the 1980s. Private organizations are poised to become major generators of improved maize germplasm in Third World countries. Both national and multinational companies are involved in maize breeding activities. Smaller companies tend to focus on seed production and distribution, often relying on the maize varieties and hybrids developed by public-sector organizations because their scales of operation are not sufficient to finance a significant research program of their own. Although private companies carefully guard their advanced inbred lines and information about pedigrees and heterotic patterns, some are willing to share portions of their germplasm collections—usually early generation maize gene pools and populations—with publicly funded international and national breeding programs for research purposes.

Outstanding Maize Germplasm Complexes

Over the past 40 years, some outstanding germplasm complexes adapted to the major tropical, subtropical, highland, and temperate maize-growing environments found in developing countries have been developed. These germplasm complexes encompass enormous genetic diversity. Vast benefits have accrued to national maize breeding programs—and to humanity—because of the widespread genetic mixing that has occurred between the elite tropical and subtropical maize gene pools of Latin America and those of Asia and Africa. The result is that today the maize genetic resources available to national maize breeding programs are much broader than they were 30 years ago.

Maize breeders in temperate zones have made less use of tropical and subtropical elite germplasm. They have been able to find enough superior genetic diversity within their elite temperate gene pools and populations, and therefore they confine most of their crossings to these materials. Biological barriers also have restricted the mixing of these diverse gene pools. The daylength sensitivity and different flowering characteristics of tropical germplasm when it is grown in high latitudes, such as the United States and Europe, has limited the use of exotic tropical materials in temperate environments. Conversely, the inadequate resistance of temperate germplasm to the various viruses, foliar diseases, and blights found in tropical and subtropical maize-growing environments has made it difficult to use temperate materials in tropical environments (Kim and Hallauer 1989).

Goodman (1985) estimated that only 1 percent of the germplasm found in U.S. commercial hybrids is from tropical gene pools. Even so, many maize breeders believe much new and useful genetic variability, for a host of characters, will be generated through greater mixing of elite temperate and tropical germplasm complexes. In recent years, private breeders in the United States have stepped up their exploratory research in the use of tropical and subtropical germplasm.

The following section describes some outstanding germplasm complexes with adaptation to tropical, subtropical, highland, and temperate environments. Almost all of these complexes have been developed by publicly funded maize research organizations.

Lowland Tropical Maize Germplasm

Wellhausen (1978) describes four outstanding Latin American elite racial complexes that have been extensively used by maize improvement research programs in developing improved genotypes for the tropics: Tuxpeño and its related Caribbean and U.S. dents, Cuban Flint, Coastal Tropical Flint, and ETO. Most improved breeding stocks in the tropics contain germplasm from these germplasm complexes, which have helped to revolutionize maize breeding in the tropics. (A series of bulletins describing these racial complexes were published in the 1950s and 1960s by the Rockefeller Foundation in collaboration with the U.S. National Academy of Sciences and the governments of Mexico, Colombia, Peru, and Brazil.)

Tuxpeño, a pure dent type, has been an exceptional racial complex. It originated on the Gulf Coast of Mexico and has a complex pedigree distinctly different from the other three races. It has influenced the development of the Cuban and West Indies dent races and is one of the putative parents of the U.S. Corn Belt dents. It combines well with ETO and the Caribbean flint complexes.

The diversity of the Tuxpeño race and various derivatives is shown in Figure 6.1. The Tuxpeño dents show high heterosis with ETO and with the Caribbean flints, suggesting that hybrid crosses using these races would be worthwhile. Recent combining-ability studies have revealed that Tuxpeño races also show high heterosis with race Yucatan. The relatively weak heterosis between Tuxpeño or the Caribbean flint groups and U.S. Corn Belt materials has been attributed to the interrelationship of these groups in the origin of the U.S. Corn Belt dents.

CIMMYT's Lowland Tropical Germplasm

CIMMYT inherited a rich stock of improved tropical and subtropical maize genetic resources from the cooperative Mexican-Rockefeller Foundation maize improvement program and from several other national research organizations. Building upon this germplasm base and drawing on many new germplasm sources, CIMMYT has developed 15 elite maize populations (and 12 corresponding gene pools) and 6 quality protein maize populations (see Chapter 7) that are adapted to the lowland tropics (Table 6.2). These lowland tropical populations were formed on the basis of genetic background, grain color (white, yellow), grain type (flint, dent), and maturity period (early, intermediate, late).

In 1984, CIMMYT initiated a series of studies of the combining ability of its various pools and populations. They were divided into eight groups, and diallel crosses were made within each group. Possible heterotic patterns for the tropical germplasm are shown in Table 6.3. Among CIMMYT's white tropical germplasm, the most promising heterosis was observed in crosses of Population 21 x Population 32, Population 21 x Population 25, Population 21 x Population 29, and Population 23 x Pool 20. Among CIMMYT's yellow tropical germplasm, the most promising crosses were Suwan-1 x Population 24, Suwan-1 x Population 27, Suwan-1 x Pool 26, and Population 26 x Pool 21. The best tropical x subtropical heterotic combinations were Population 32 x Population 44 (American Early Dent-Tuxpeño), Population 43 x Population 42 (ETO-Illinois), and Population 43 x Population 44.

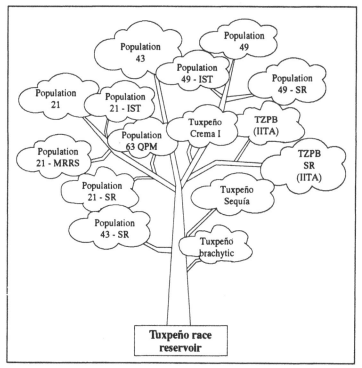

FIGURE 6.1 The Tuxpeño family tree (*Source:* CIMMYT 1992).

TABLE 6.2 CIMMYT's advanced lowland tropical normal and quality protein maize populations.

Population[a]	Type[b]
21 Tuxpeño-1	LWD
22 Mezcla Tropical Blanca	LWDF
23 Blanco Cristalino-1	IWF
24 Antigua-Veracruz 181	LYD
25 Blanco Cristalino-3	LWF
26 Mezcla Amarilla	IYFD
27 Amarillo Cristalino-1	IYF
28 Amarillo Dentado	LYD
29 Tuxpeño Caribe	LWD
30 Blanco Cristalino-2	EWF
31 Amarillo Cristalino-2	EYF
35 Antigua-Rep. Dominicana	IYD
36 Cogollero	LYD
43 La Posta	LWD
49 Blanco Dentado-2	IWD
61 Early Yellow Flint QPM	EYF
62 White Flint QPM	LWF
63 Blanco Dentado-1 QPM	LWD
64 Blanco Dentado-2 QPM	LWD
65 Yellow Flint QPM	LYF
66 Yellow Dent QPM	LYD

[a] QPM = quality protein maize.
[b] E = early maturity. I = intermediate maturity. L = late maturity. D = dent. F = flint. W = white. Y = yellow.
Source: CIMMYT.

TABLE 6.3 Possible heterotic combinations with CIMMYT's tropical maize populations.

Population	Possible heterotic partner
21 Tuxpeño 1	Pops. 32 and 25 and Pool 23
22 Mezcla Tropical Blanca	Pop. 32
23 Blanco Cristalino-1	Pool 20
24 Antigua-Veracruz 181	Pop. 36 and Suwan-1
25 Blanco Cristalino-3	Pop. 21
26 Mezcla Amarillo	Pool 21
27 Amarillo Cristalino-1	Pool 25, Suwan-1, Pop. 44*
28 Amarillo Dentado	Pop. 24 and Suwan-1
29 Tuxpeño Caribe	Pop. 32
31 Amarillo Cristalino-2	Pop. 49*
32 ETO Blanco	Pops. 21, 22, 29, 44
36 Cogollero	Pop. 24
43 La Posta	Pops. 42 and 44
49 Blanco Dentado-2	Pops. 26* 31*; Pool 21*

* Material has different grain color than the matching population in the left column.
Source: CIMMYT 1987a.

IITA's Lowland Tropical Germplasm

The first high-yielding populations with resistance to maize streak virus were TZSR-W-1 and TZSR-Y-1, two late-maturing, white and yellow semi-flint materials developed by IITA. Since then, IITA has developed a range of tropical maize populations of different maturity and grain color and hardness (Table 6.4).

IITA has developed dozens of outstanding inbred lines for tropical areas that are made available to maize researchers in public and private organizations through two international trials. In addition to these officially released inbred lines, 20 inbreds with lowland tropical adaptation have been selected based on heterotic patterns and resistance to diseases, insects, and striga. Seed of these inbreds is available from IITA.

Suwan-1 Tropical Germplasm

The open-pollinated variety, Suwan-1, developed in Thailand and based on elite Caribbean yellow flint germplasm, is an example of the benefits from international cooperation in germplasm exchange and maize research. The Suwan-1 story in Thailand began in 1960 with the introduction of the lowland tropical variety Tiquisate Golden Yellow, which had been released in Guatemala in 1954. Tiquisate Golden Yellow is a mixture of two strains of maize of Cuban origin: a white semi-dent and a golden yellow flint. The variety was well adapted to Thailand, and its golden yellow semi-flint grain type was in demand in international markets. Farmers quickly adopted this variety, which was called Guatemala. In 1967, CIMMYT supplied the Thai national program with Caribbean tropical flint and dent (Tuxpeño) materials similar

TABLE 6.4 IITA's tropical maize populations.

Population	Developed from	Type[a]
TZB	Nigerian Composite B, Latin America germplasm	SLWF
TZB-SR	TZB, Reunion, IB32	SLWF-SR
TZPB	Tuxpeño Crema 1, C7	FoLWD
TZPB-SR	TZPB, Reunion, IB32	FoLWD-SR
TZSR-W-1	IB32, Reunion, TZB, TZPB, some CIMMYT germplasm	SLWF-SR
TZSR-Y-1	IB32, Reunion, Local Yellow, some CIMMYT germplasm	FoLYF-SR
TZUT-W	TZSR-W-1, U.S. germplasm	SIWD-SR
TZUT-Y	TZSR-Y-1, U.S. germplasm	SIYD-SR
TZESR-W	Early varieties-Burkina Faso, IB32, Reunion	SEWF-SR
TZESR-Y	Early varieties-Burkina Faso, IB32, Reunion	SEYD-SR
TZMSR-W	East Africa germplasm, IB32, Reunion	MLWF-SR

[a] E = early maturity. I = intermediate maturity. L = late maturity. W = white. Y = yellow. F = flint. D = dent. S = savanna. Fo = tropical forest. M = mid-altitude. SR = streak resistant.
Source: IITA.

to Tiquisate Golden Yellow in adaptation and grain type. Using this collection of germplasm, Thai breeders selected 36 germplasm complexes for compositing. After four cycles of recurrent selection, this population yielded 20 to 30 percent more than Tiquisate Golden Yellow, and it was officially named Thai Composite #1.

By the early 1970s, however, downy mildew disease had invaded Thailand, and Thai maize varieties, including Thai Composite #1, were highly susceptible. To overcome this susceptibility, DMR 1 and DMR 5, Philippine sources of downy mildew resistance, were backcrossed three times to Thai Composite #1. The new material, called Thai Composite #1 DMR, was released as Suwan-1 in 1974. In addition, 250 of the earliest-maturing families of this population were used to create Thai Composite #1 Early DMR, which was released as the early maturing variety, Suwan-2. Suwan-1 has spread across Thailand and has been used as a breeding material in many other countries, both as a direct release open-pollinated variety and for developing inbred lines that have been crossed with elite inbred lines from the United States, Brazil, the Philippines, and elsewhere to produce excellent three-way and double-cross hybrids adapted to tropical environments.

Indian Tropical Germplasm

During the 1950s, some outstanding elite germplasm complexes were introduced into Indian maize breeding programs to broaden the germplasm base. In particular, germplasm from Latin America (Colombia, Peru, Venezuela, Antigua, Cuba) and from southeastern United States (North Carolina and Florida) was used in the development of inbred lines. Inbred lines were developed from these materials and were used to develop the double-cross hybrids Ganga 1, Ganga 101, Ranjit, and Deccan, all of which were dent x flint crosses.

During the 1960s, the All-India Coordinated Maize Improvement Program further developed a number of improved composites, hybrids, and open-pollinated varieties based upon new tropical and subtropical germplasm complexes. Outstanding among these populations were J-1 (also known as Naraingarh Complex) and Sona (J1 x 11J). In 1967, six high-yielding composites—Amber, Vijay, Jawahar, Kisan, Sona, and Vikram—were developed from these populations and released as open-pollinated varieties.

Brazilian Tropical Germplasm

Much of Brazil's maize growing areas are subtropical to tropical. Outstanding germplasm complexes have been developed from the local Cateto race, which encompasses a group of yellow to orange flint maizes occurring in Brazil and assumed to have been grown by indigenous people living along the Atlantic coast from Brazil to Guyana. Although not a high-yielding material itself, Cateto exhibits high combining ability when crossed with many races. The flint population ESALQ-VF-1 is based on Cateto, ETO, and other Caribbean flints and has been widely used to develop open-pollinated varieties and inbred lines.

Cateto germplasm combines especially well with Tuxpeño germplasm. Tuxpeño has had a striking effect in increasing the yield potential of hybrids based on local materials. The dent population ESALQ-2-VD-2 is based mainly on Tuxpeño germplasm has been widely used in the development of open-pollinated varieties and inbred lines.

Brazil's Tuxpeño and Tuxpeño x Stiff Stalk Synthetic germplasm complexes combine well with the Cateto Sulino and Cateto Assis Brasil populations, ETO, Caribbean flints (such as Suwan-1), and Rio Grande do Sur dent materials (Paulista dents). Tuxpeño x Suwan-1 also is being used in new hybrid combinations.

Brazilian germplasm that has tolerance to aluminum toxicity, developed at EMBRAPA's national maize and sorghum research center in Sete Lagoas is wide grown in Cerrado soils and is being used extensively wherever acid soils with aluminum toxicity are an important constraint to maize cultivation. More detail about Brazilian maize research and germplasm is given in Chapter 10.

Other Tropical Germplasm

Some other examples of germplasm sources that are widely used by breeders in developing superior genotypes for lowland tropical areas are Venezuela-1, an orange flint open-pollinated variety based on Caribbean germplasm; the semi-dent hybrid Obregon developed in Venezuela and based on inbred lines from ETO and other Caribbean flints and crossed with inbred lines derived from Tuxpeño; La Maquina, a white dent open-pollinated variety developed in Guatemala from Tuxpeño materials; Katumani Composite, a white dent open-pollinated variety developed in Kenya that exhibits fair drought resistance; and the Peruvian yellow hybrid PM 701, developed by the University of Agriculture at La Molina, based on inbred lines derived from Perla, a local flint variety crossed with Cuban flint lines.

Improved Maize Germplasm for Tropical Mid-Altitudes

Improved germplasm for intermediate-elevation subtropical maize environments (900 to 1,700 meters elevation), which are found mainly in Africa, requires resistance to maize streak virus, common rust, northern virus, which is confined to Africa, resistance to these diseases also is essential in Latin America and Asia.

Zimbabwe's Intermediate-Elevation Germplasm

Some of the best germplasm for the intermediate elevations of the tropics has been developed in Zimbabwe. The national maize program has produced a series of outstanding open-pollinated varieties and hybrids since the 1930s. The breeding program was based on the varieties Southern Cross, Salisbury White, and to a lesser extent, the U.S. variety Hickory King (Tuxpeño background), which arrived in southern Africa via traders in the 19th century.

In 1960, the world's first commercial single-cross hybrid, SR 52, was released in Zimbabwe. SR 52 is based on lines derived from the important heterotic groups Salisbury White and Southern Cross. Both populations are high-yielding and well-adapted, and the initial inbreds selected from this material were outstanding. Some of the early selections are still in use today and recycling of existing elite inbred lines and backcrossing are common sources of new inbred lines. New hybrids based on recycled SR52 inbred lines started to replace the old SR52 in the 1980s. Two shorter-season hybrids, R201 and R200, were grown in 1991 on about 70 percent of the hybrid area in Zimbabwe.

CIMMYT's Mid-Altitude Tropical Germplasm

In 1985, CIMMYT established a maize breeding program at the University of Harare, Zimbabwe, to develop a broader range of improved germplasm for the intermediate elevations of tropical Africa. A major breeding objective has been to incorporate resistance to maize streak virus and other diseases of importance. The program focuses on development of open-pollinated varieties and synthetics as well as development of inbred lines and hybrid combinations. Some elite inbred lines having resistance to streak virus and also rust and leaf blight have been developed.

IITA's Mid-Altitude Tropical Germplasm

Two IITA populations, TZMSR-W and TZEMSR-W, are serving as good sources for intermediate-elevation germplasm development for sub-Saharan Africa. TZMSR-W is a late-maturing, white-grain, semi-flint population adapted to elevations of 1,000 to 1,600 meters. TZEMSR-W is an early maturing, white-grain population with good adaptation in intermediate-elevation environments in Nigeria, Cameroon, and Zaire. The IITA collaborative program with Cameroonian national and Zaire. The IITA collaborative program with Cameroonian national maize research organizations has released some outstanding inbred lines for intermediate elevations.

Subtropical Germplasm

Maize was domesticated in the tropics and gradually moved to higher latitudes over the ages. The subtropics is a transitional zone between the tropics and temperate zones.

The subtropical germplasm in various national breeding programs has been derived from mixing local landraces with tropical and temperate germplasm. Depending on the season and elevation in which maize is grown in the subtropics, the climate can either be tropical or temperate. In the summer months, production environments in lowland subtropical areas are basically tropical in nature—hot days and nights. In the winter months, production environments in lowland subtropical areas can be more temperate—warm to hot days, cooler nights. These different agroecological conditions affect which germplasm sources have been most suitable to national breeding programs.

CIMMYT's Subtropical Germplasm

CIMMYT has various gene pools and populations that have adaptation to subtropical environments. Its advanced populations are composed of mixtures of outstanding temperate and tropical germplasm (Table 6.5). In recent years, the levels of resistance to northern leaf blight and common rust has been raised substantially in CIMMYT germplasm.

Diallel studies of subtropical germplasm done by CIMMYT scientists has provided information on the combining ability of these different materials. Among CIMMYT white- and yellow-grain, early maturing, subtropical populations, Population 48 (Compuesto de Hungria) and Pool 30 (subtropical-temperate early yellow dent) showed high general combining ability. Population 48 also performed well in crosses with CIMMYT Pools, 27, 28, and 30 and with Population 46 (Templado Amarillo Cristalino).

In the diallel containing CIMMYT's intermediate-maturity subtropical populations (white and yellow), Population 42 (ETO-Illinois) and Population 47 (Templado Blanco Dentado) showed good combining ability effects for yield. Table 6.6 summarizes the possible heterotic combinations with CIMMYT populations.

Two new subtropical populations have recently been developed at CIMMYT: Siwdent-1 (subtropical intermediate white dent) Population 500 and Slawdent-1 (subtropical late white dent) Population 600. These populations have been developed from crosses between S_1 lines derived from commercial Mexican hybrids and some advanced lines from CIMMYT's subtropical pools and populations that have resistance to northern leaf blight and tarspot. Selection emphasis is on yield, resistance to northern leaf blight and common rust, reduced plant height, and resistance to lodging.

Indian Subtropical Germplasm

Indian federal and state maize research programs have developed germplasm especially for the winter (which has a temperate to subtropical environment) in the northern zones of the country (Punjab, Uttar Pradesh, Bihar). The composite Partap, developed from crosses among three hybrids—Vijay x Pioneer 102, Pioneer 114 x Vijay, and Ganga 5 x Pioneer 104—showed good cold tolerance and was improved through a recurrent selection program. Several open-pollinated varieties, under the name of Partap, have been released since the early 1980s. The latest version, Partap C_5, has been released for the Indo-Gangetic plains. Lakshmi, developed from CIMMYT Population 44 (AED-Tuxpeño), is a widely grown open-pollinated variety for winter maize production in Bihar. The most popular hybrids for winter maize production are Ganga Safed 2 and High Starch, two nonconventional hybrids involving a single-cross based on U.S. and Caribbean germplasm that was crossed with Rudrapur Local, a local variety, to produce a flint hybrid and with Jellicorse synthetic to produce a dent hybrid.

TABLE 6.5 CIMMYT's advanced subtropical maize populations.

Population[a]	Type[b]
32 ETO Blanco	IWF
33 Amarillo Subtropical	IYF
34 Blanco Subtropical	LWF
42 ETO-Illinois	LWD
44 Amarillo Early Dent-Tuxpeño	LWD
45 Amarillo Bajio	IYD
46 Templado Amarillo Cristalino	EYF
47 Templado Blanco Dentado-2	IWD
48 Compuesto de Hungria QPM	IWD
67 Templado Blanco Cristalino QPM	IWF
68 Templado Blanco Dentado QPM	IWD
69 Templado Amarillo QPM	IYF
70 Templado Amarillo Dentado QPM	IYD

[a] QPM = quality protein maize.
[b] E = early maturity. I = intermediate maturity. L = late maturity. D = dent. F = flint. W = white. Y = yellow.
Source: CIMMYT.

United States Subtropical Germplasm

In the United States, several maize research programs have made crosses of temperate and tropical inbred lines to use as sources for developing new subtropical and tropical inbred lines. The University of Hawaii pioneered efforts to convert U.S. inbred lines to forms adapted to tropical and subtropical environments. Lines have been developed with resistance to several major pests and diseases, northern leaf blight, maize mosaic virus I, common rust, and maize earworm. A major shortcoming of these inbred lines is disease resistance. They often are susceptible to the more virulent forms of foliar diseases and blight found in subtropical and tropical areas (Brewbaker, Logrono, and Kim 1989). These lines are distributed worldwide under the Maize Inbred Resistance (MIR) program coordinated by the University of Hawaii.

Maize breeders at the University of Florida have developed a series of subtropical inbreds by crossing temperate with tropical germplasm (Gracen 1986). Several have shown good disease resistance and adaptation in tropical or subtropical environments or both.

Highland Germplasm

Highland areas can be subdivided into tropical highland transition zones, tropical highlands, and temperate highlands. Most of the Third World's highland maize area is found in tropical and subtropical latitudes between the equator and about 30° north and south.

Mexican Highland Germplasm

The Mexican highland landraces and racial complexes initially identified by the scientists in the Mexican government-Rockefeller maize improvement program that were used in subsequent breeding belong to four racial groups: Celaya, Chalqueño, Conico Norteño, and Conico.

TABLE 6.6 Possible heterotic combinations with CIMMYT's subtropical maize populations.

Population	Possible heterotic partner
33 Amarillo Subtropical	Pop. 45
34 Blanco Subtropical	Pop. 42; Pool 34*
42 ETO-Illinois	Pops. 34, 43, 45*, 47
44 AED-Tuxpeño	Pops. 32, 25, 27*, 43
45 Amarillo Bajio	Pop 33; Pool 33
46 Templado Amarillo Cristalino	Pop. 48; Pool 30
47 Templado Blanco Dentado-2	Pop. 42
48 Compuesto de Hungria	Pop. 46; Pools 27*, 28*, 30

*Different grain color than the population in the first column.
Source: CIMMYT 1987a.

The high-elevation (2,200 to 2,400 m) long-season hybrids H-129 and H-133 were developed from lines derived from the Chalqueño racial groups. The intermediate-maturity hybrid H-28, which is adapted to 1,800 to 2,200 meters elevation, was developed from inbred lines derived from the Celaya racial group. The racial group Conico Norteño was used to develop the population Montaña, which when crossed with [Sabanero (floury) x Pira Naranja (popcorn)] became the parent population for the variety Ecuador 573, which proved useful in the development of the Kenyan hybrids.

Andean Region Highland Floury Germplasm

In 1977, CIMMYT posted a maize breeder in Quito, Ecuador, to work with INIAP researchers to improve the soft, large-kernel floury and morocho (large grain, harder endosperm) types preferred in the Andean highlands. To broaden the genetic base of these grain types, four floury and four morocho gene pools were formed using elite races, cultivars, and improved varieties from the Andean region and other highland germplasm, mainly from Mexico and Guatemala, held in CIMMYT's maize seed bank. The Mexican material Compuesto Cacahuacintle has produced some excellent early maturing, white floury-grain varieties that are being grown in Ecuador (INIAP 101), Bolivia, and Peru. The racial group Chalqueño has also been used extensively. INIAP/CIMMYT pools 1 and 4 formed good sources for floury varieties and pools 7 and 8 for morocho varieties.

The yellow morocho variety INIAP 180, developed from ICA-V-507 and MB-517 x ICA-V-507, has been an outstanding material. Other superior materials are INIAP 130, an early yellow floury germplasm; INIAP 176; and INIAP 178, which has INIAP 176 in its parentage.

Transition-zone Highland Germplasm

In 1985, using materials developed in Tanzania in the 1970s, CIMMYT began the formation and improvement of a new germplasm complex for transition zones found in eastern and southern Africa that require germplasm different from that adapted to intermediate elevations. The modified version of CIMMYT Pool 9, called Pool 9A, is a late-maturity, white semi-dent population, is based largely on Kitale Synthetic II, Ecuador 573, and SR 52 germplasm. Other important Pool 9A germplasm sources include highland Tuxpeños from Mexico and Guatemala, and Montaña from Colombia. Pool 9A has performed extremely well in highland transition zones, especially in sub-Saharan Africa.

Another germplasm complex for the transition zone is being developed through the IITA/IRA/NCRE cooperative maize improvement program in Cameroon. It consists of germplasm from CIMMYT (Pool 9A), Kenya (highland hybrids), Guatemala (V301, V304), IITA (TZMSR), United States (DeKalb 690), and Zimbabwe (SR52, ZS206).

Kitale Highland Germplasm

During the 1950s, a maize breeding program was established at Kitale Research Station, Kenya, involving the United States government and the East African Agriculture and Forestry Research Organization. Researchers gathered the best local maize varieties from farmers' fields and assembled them into several breeding populations. This germplasm can be traced back to Hickory King, which reached southern Africa in the 19th Century and then over generations was selected for adaptation as the crop moved slowly north. Because of the material's narrow genetic base, initial efforts to improve it met with little success. In 1957, exotic germplasm was introduced from the Colombian-Rockefeller Foundation maize improvement program, notably Ecuador 573 and Costa Rica 76. When crossed with local populations, these Latin American varieties had a dramatic impact on yield potential. The outstanding germplasm complexes, Kitale I and II, were developed by intercrossing local breeding material with materials from Colombia. A variety-cross hybrid, H 611 (KII x Ecuador573), was also released for commercial use.

The varieties Kenya Flat White and Ecuador 573 turned out to be especially important germplasm sources for the development of inbred lines. Hybrids and varieties based upon inbred lines from these maize populations are extensively grown in Kenya and its neighbors.

Temperate-zone Highland Germplasm

CIMMYT has developed temperate highland populations that have white and yellow semi-dent grain by intercrossing improved tropical highland germplasm with northern U.S. and European materials. These materials carry cold tolerance, earliness, resistance to northern leaf blight and common rust, and other important attributes. Inbred lines are being developed from these materials. Photoperiod insensitivity in these germplasm complexes is being developed through collaborative research with Asian countries that have temperate highland environments. Two early maturing populations (Pops. 800 and 845) have been formed, and two intermediate-maturing populations are being formed.

Temperate Germplasm

Maize-growing areas in the United States, Canada, Argentina, Chile, Europe, the Middle East, and East Asia have temperate environments.

The germplasm in use is not as genetically diverse as tropical and sub-tropical germplasm. Because the temperate maize-growing environments tend to be relatively uniform, they permit selection of highly adapted hybrids. Varieties that have uniformity in plant and ear height and in maturity also are desirable for mechanized maize production, which is prevalent in temperate regions.

U.S. Corn Belt

Corn Belt maize hybrids are derived primarily from inbred lines developed from Reid Yellow Dent and Lancaster Sure Crop varieties. Iowa Stiff Stalk Synthetic inbred lines (Reid germplasm) crossed to Lancaster Sure Crop inbred lines remain the predominant heterotic combinations in the United States. In 1985, nearly 90 percent of all U.S. Corn Belt hybrids contained Reid germplasm in their pedigrees and 13 percent contained Lancaster germplasm (Darrah and Zuber 1986). Most European, Chinese, Argentinian, and Chilean maize breeding programs also used Lancaster- or Reid-derived inbred lines, or both, in their hybrid development programs. Many inbred lines have been developed from these two germplasm groups (Table 6.7).

Maize research organizations in developing countries, such as China, North Korea, Argentina, Chile, and Yugoslavia, that have temperate maize-growing conditions have made extensive use of U.S. inbred lines such as B37, B73, C103, Oh43, and Mo17 for hybrid development. U.S. agricultural universities in Hawaii and Florida—and IITA in West Africa—have helped to develop versions of these inbreds with added resistance to subtropical and tropical diseases and with better photoperiod adaptation. Another line of work that is being followed at Iowa

TABLE 6.7 Important public U.S. inbred lines based on Reid and Lancaster germplasm complexes.

Inbred	Year released	Source	Maximum use[a] (%)
C103	1949	Lancaster Sure Crop	12
Oh43	1949	Lancaster Sure Crop	16
B14	1953	Reid (Stiff Stalk Syn.)	9
W64A	1954	Reid (Stiff Stalk Syn.)	13
B37	1958	Reid (Stiff Stalk Syn.)	26
A632	1964	Reid (Stiff Stalk Syn.)	15
Mo17	1964	Lancaster Sure Crop	12
B73	1972	Reid (Stiff Stalk Syn.)	16
B84	1978	Reid (Stiff Stalk Syn.)	<1
B94	1992	Reid (Stiff Stalk Syn.)	<1

[a] Percent of total crop planted using indicated inbred line as one parent.

Source: Darrah and Zuber 1986; and authors' data.

State University is to change some CIMMYT populations to temperate adaptation. For example, Tuxpeño, ETO, and Antigua are now becoming well adapted to Iowa conditions.

Europe

A wide range of maize germplasm sources have been introduced into Europe from the Western Hemisphere. Trifunovic (1978) classifies these diverse germplasm complexes into four groups: Corn Belt dent-like forms, Southeastern European flints, northern European and Alpine flints, and southern European or Mediterranean types (mostly flints).

A common heterotic group in Europe is based on U.S. early dents x European cold-tolerant flint lines. A widely grown French single-cross hybrid, F2 x F7, was derived from the same population. Initially, U.S. inbred lines and hybrids from Wisconsin and Minnesota were introduced into Europe. Later INRA, the French agricultural research institute, developed several inbred lines from the European flint pool with earlier maturity and greater cold tolerance than the U.S. germplasm. INRA 258, the first double-cross hybrid developed by European breeders, was released in 1958 and was based on a U.S. Corn Belt single cross (dent) and a European early maturing, cold-tolerant, single cross (flint). This hybrid helped push maize production further north in Europe. Subsequently, the early maturing three-way cross hybrid, LG-11, released by the French cooperative LimaGrain in 1970 (based on work done by INRA), became the most important variety in northern Europe (McMullen 1987).

China

Most farmers in China grow hybrids based on U.S. germplasm x Chinese flints. The original double-cross hybrids introduced during the 1950s and 1960s were based on elite Chinese flint germplasm crossed with U.S. Corn Belt inbred lines. Chinese open-pollinated varieties that have produced outstanding varieties include Rii28, Houbie, and Dahuan 46. U.S. Corn Belt inbred lines initially included 38-11, L289, W19, W20, W24, and M14. More recently, U.S. Corn Belt inbreds such as B37, Oh43, Mo17, and A619Ht have proven to be useful. At present, single-cross hybrids based upon Mo17 x Chinese flint inbred lines are the most popular and highest-yielding genotypes in China.

Southern Cone of South America

Argentina and Chile have temperate environments in which about 2 million hectares of maize are grown. Argentina's farmers plant some of this area with hybrids developed from local Cateto flints with a dark orange grain color. They also make use of U.S. Corn Belt germplasm where intermediate maturity is needed and of European germplasm where early maturity is needed.

7

Breeding Advances in Tropical and Subtropical Maize

The aim of maize improvement research is to produce a continuing stream of improved genotypes that are highly efficient in their use of soil fertility, water, and light in spite of the biotic and abiotic stresses that can harm the crop. In the past three decades, national and international maize research programs have made great progress in improving the agronomic traits of elite maize germplasm complexes and increasing the genetic resistance or tolerance to a host of environmental stresses.

Improved Harvest Index

Indigenous tropical maize germplasm is inherently inefficient because of its tall stature, abundant foliage, large tassels, and, hence, its relatively low harvest index (ratio of grain to total aboveground dry matter). From an evolutionary standpoint, a larger, leafier plant probably had better chances of surviving the ravages of tropical diseases and pests and of competing with aggressive weed species for sunlight during early vegetative growth. Most indigenous tropical germplasm sources are over 3 meters high and have harvest indexes of only 30 percent (compared with 50 to 55% for U.S. Corn Belt hybrids). They do not tolerate high plant density and fertilizer nutrient input.

The characteristics of the high-yielding semidwarf wheat and rice varieties and of hybrid maize in U.S. Corn Belt, clearly demonstrates the benefits of reducing plant height in many tropical maize environments. Physiologically, shorter plant types partition more of their total biomass to grain than to stover, and permit cultivation of higher density plant stands (resulting in higher yields) than do tall materials. Shorter plant types also are more easily managed by farmers and provide new intercropping alternatives because they permit greater light penetration into the plant canopy.

Maize breeders have explored various ways to shorten tropical maize. During the 1970s, CIMMYT conducted a special research project in tropical materials to determine the optimum plant type for grain production (Johnson et al. 1986). Because the major dwarfing genes in maize are associated with some undesirable morphological traits, the main methodology breeders employed was recurrent selection for shorter plant height within high-yielding widely adapted germplasm. The population chosen was Tuxpeño Crema 1, a white dent grain, lowland tropical maize material. During 20 cycles of recurrent selection, many changes occurred in this population.

Table 7.1 compares these changes over 17 cycles of selection. Plant height was reduced by nearly 50 percent, harvest index increased from 30 percent to 47 percent, grain yield at optimum planting densities (higher density as plants were shortened) increased by more than 50 percent, and the population maturity became 11 days shorter. For maximum grain yield in most tropical environments, materials from cycles 12 to 15 are optimum plant types (175 to 190 cm height). The study also demonstrated the high correlation between shorter plant height and earlier maturity. CIMMYT's Population 49, an intermediate to early material, is based upon cycle 17 of Tuxpeño Crema 1.

Some maize researchers have used dwarfing genes to reduce plant height. Brachytic-2 (*br-2*) has been the most promising dwarfing gene, even though it is associated with undesirable traits such as very broad leaves, reduced yield, and, in some cases, excessively short plants. These effects led many maize researchers to abandon early efforts to shorten tropical maize with major genes. Brazilian researchers, however, carried on the work using two brachytic germplasm sources (a white Tuxpeño and an orange flint) to reduce the height in two tropical tall populations—ESALQ-VD-2, a yellow dent material based largely on Tuxpeño germplasm, and ESALQ-VF-1, an orange flint material made up mainly of Cateto and Caribbean germplasm. Inbred brachytic lines

TABLE 7.1 Comparison of various cycles of selection for reduced plant height in Tuxpeño Crema 1, when grown near optimum plant density.

Cycle of selection	Plant height (cm)	Days to maturity	Grain yield[a] (t/ha)	Total dry matter (t/ha)	Harvest index (%)
0	273	125	4.05	14.9	30
6	211	122	5.54	14.7	38
9	203	122	5.67	15.3	39
12	196	119	6.18	15.4	41
15	173	116	6.73	15.1	46
17	156	114	6.23	13.1	47

[a] Mean yields for 2 years in Mexico.
Source: CIMMYT.

developed from these populations have been used to produce commercial hybrids in Brazil (Paterniani 1990). Chinese scientists, particularly in southern China, have also made progress in developing brachytic lines, and several brachytic hybrids are now in commercial use.

In addition to selecting for shorter plants to improve harvest-index, breeders have selected for other traits such as reduced tassel size. During the 1970s, CIMMYT carried out a recurrent selection program for reduced tassel size and leaf area on three elite populations: Pop. 21 (Tuxpeño-1), Pop. 35 (Antigua x Republica Domincana), and Pop. 32 (ETO Blanco). Over the six cycles of selection in these populations, significant increases in grain yield, harvest index, and optimum plant density were obtained (Fischer, Edmeades, and Johnson 1987). In Brazil, the work of Geraldi, Biranda, and Vencovsky (1985) in mass selection for smaller tassels in three maize populations showed that grain yield increased as tassel size (number of branches) was reduced.

The possibility of developing of shorter, more grain-efficient tropical maize materials—mainly through recurrent selection schemes to accumulate polygenes—has been amply demonstrated and has become a standard maize improvement objective in tropical maize breeding programs. As a result, the harvest index of most improved tropical maize materials has increased from about 30 percent to 40 to 45 percent, with plant height ranging from 2.0 to 2.5 meters. Much of the improvement in yield potential of these short plant types has been due to their ability to respond to higher plant density without lodging or an increase in barrenness.

Disease Resistance

Warm year-round temperatures and ample moisture are conducive to the spread of maize diseases. Many virulent tropical diseases tend to occur simultaneously in certain environments. Southern rust and southern leaf blight, for example, often break out together in humid lowland tropical environments. Similarly, common rust and northern leaf blight thrive in subtropical (and mid-altitude tropical) environments with cool humid climates. Stalk and ear rots are widespread diseases of maize in most maize-growing environments.*

Considerable progress has been made in the past 20 years in developing host plant resistance to the major diseases of tropical maize. In general terms, disease resistance in maize is controlled (1) by one or a few genes, (2) by many genes, or (3) by the cytoplasm. For most maize dis-

---

* A list of major maize diseases and their causal organisms can be found in Table 3.3.

eases, the inheritance of resistance involves many genes (polygenic resistance) and is therefore complex and controlled by additive action of genes. To accumulate a greater number of desirable genes for durable disease resistance, some form of recurrent selection is needed. If properly carried out, recurrent breeding methodologies can increase gene frequency for the desired character without causing detrimental changes in other desirable agronomic traits.

For a few maize diseases, inheritance of resistance is not complex. That is, one or two major genes provide resistance. When such a source of resistance is located, transferring it into advanced breeding material through a backcrossing program is rather routine. The most durable resistance to tropical diseases, however, is polygenic (race-nonspecific) and generally should be pursued over single-gene (race-specific) disease resistance, which often is precarious and nondurable.

High-yielding lowland tropical germplasm sources have been developed with polygenic resistance to southern rust, stalk and ear rots, downy mildew, maize streak virus, maize stunt, and southern and northern leaf blight. Single-gene resistance also has been exploited in controlling common rust and northern leaf blight. However, such single-gene resistance has tended to be relatively short-lived.

Improved subtropical germplasm—often developed using temperate materials—has not equaled the disease resistance of many tropical materials. Except for some tolerance of common rust, southern leaf blight, and northern leaf blight, most temperate inbreds are highly susceptible to tropical diseases, primarily because temperate zone germplasm complexes have not been selected for resistance to tropical diseases. During the past decade, however, maize scientists have been able to raise the disease resistance of high-yielding subtropical materials. This has been achieved by introgressing more tropical germplasm into subtropical materials. The intermixing of elite tropical and subtropical germplasm has not only incorporated new sources of resistance to leaf blights, rust, stalk and ear rots, and smuts, but it also has led to greater yield potential.

Rusts

Three major types of rusts can attack maize: southern rust, common rust, and tropical rust (caused by *Physopella zeae*). Southern rust and common rust are the most prevalent, and they cause the greatest economic damage. These two species of rust are incited by differing climatic factors. Southern rust prevails in subtropical and highland environments and common rust prevails in the lowland tropics and subtropics.

Several races of southern rust exist. The causal organism, *Puccinia polysora*, was confined to the Americas until 1949 when it was introduced into West Africa. Within 10 years it had spread to all maize-growing areas of the continent. For 2 to 3 years, a severe southern rust epidemic gripped western and central Africa. The introduction of Tuxpeño and other rust-resistant improved maize germplasm from Mexico and Latin America, led to the eventual replacement of local West African varieties by Mexican and Caribbean germplasm.

As many as 11 specific genes—designated as *Rpp1* through *Rpp11*—exist for resistance to southern rust in maize (Renfro unpublished). As a result of natural selection pressure and the tropical maize breeding programs that have been conducted for several decades, most of the Mexican dents and Caribbean dents and flints are highly resistant to southern rust. Virtually all of CIMMYT's tropical lowland germplasm has moderate to high field resistance, with the late and intermediate maturity germplasm showing the best resistance. Tuxpeño germplasm in particular has good polygenic resistance.

Common rust can be found in subtropical maize-growing environments and in the relatively cooler environments (17 to 25°C) in the tropical mid-altitudes and highlands. Improved highland germplasm based upon Mexican and Andean landraces from high elevations (2,200 to 2,600 meters) is resistant to common rust. Good polygenic (horizontal) resistance to common rust exists in most improved subtropical germplasm. This resistance is conferred by numerous genes.

Subtropical populations developed by crossing temperate and tropical germplasm have some resistance to common rust that is conferred in large part by the tropical germplasm (although higher levels of resistance are generally still needed). All of CIMMYT's normal subtropical pools and populations, the Subtropical Multiple Disease Resistance (MDR) Population, and its highland tropical populations now have moderate to high resistance to common rust. Good progress has been made to develop the resistance to common rust in CIMMYT's subtropical quality protein maize. Resistance to common rust is low in most South Asian subtropical germplasm, which has considerable proportions of susceptible temperate germplasm in its background.

Most temperate germplasm has lacked resistance to common rust, although some U.S. temperate inbreds lines (Oh545, Oh515, Pa762, and C103) carry general resistance genes (Kim 1990).

Leaf Blights

Along with the rusts, leaf blights are common and ubiquitous leaf fungi that attack maize. Depending on temperature and climate, leaf blight is incited by *Helminthosporium maydis* in the warmer lowland tropics and *Exerohilum turcicum* (formerly called *Helminthosporium turcicum*) in the cooler climates of the intermediate elevations and highlands and in the winter seasons of tropical lowlands.

Southern Leaf Blight. Southern leaf blight (caused by *Helminthosporium maydis, Cochliobolus hetrosphus, Bipolaris maydis, Drechslera maydis*) is one of the most ubiquitous maize diseases. This fungus has three major races that attack maize. Race O occurs mainly in subtropical and tropical areas; race T occurs in temperate areas and is highly virulent to maize having the Texas-type male sterile cytoplasm; race C, recently identified in China, attacks maize that has C cytoplasm.

Southern leaf blight race T, first reported in the Philippines in 1959, reached epidemic proportions in the United States in 1970. It attacked all commercial maize hybrids that had Texas male-sterile cytoplasm, or 85 percent of all hybrids then in use in the United States The epidemic cause more than $1 billion in crop losses. Normal cytoplasm maize is moderately resistant to races T and C. The recessive gene, *rhm*, confers resistance to race O until the flowering stage of plant development. Afterwards, it continues to have a positive effect on polygenic resistance even though its effectiveness declines. The North Carolina State University inbred line NC 250, developed from the cross (Nigeria composite A-rb x B 37) x B 37, is reported to carry the *rhm* gene (Renfro 1985).

Many tropical germplasm complexes have good resistance to southern leaf blight. In addition, many tropical inbred lines developed at CIMMYT and IITA are highly resistant. Subtropical materials are more susceptible. Among CIMMYT subtropical germplasm, Population 42 (ETO-Illinois) and Population 44 (AED-Tuxpeño) were the most resistant.

Several subtropical to temperate inbred lines developed at the University of Florida carry good levels of resistance to southern leaf blight. The U.S. Corn Belt line, Mo 17, also shows good tolerance. Conversely, some widely used U.S. inbred lines, such as B73 (Iowa) and F44 (Florida), are highly susceptible to southern leaf blight.

Northern Leaf Blight. Northern leaf blight is common in subtropical and highland environments and in winter-season maize grown in tropical environments. Susceptible varieties can suffer severe outbreaks in subtropical and highland environments that have cool temperatures and heavy and frequent dew. Northern leaf blight resistance is less important in tropical germplasm grown during the main (summer) growing season because high temperatures do not favor disease development. Genes exist both for quantitative and qualitative resistance to *Exerohilum turcicum*. Major genes for resistance are *Ht1, Ht2, Ht3,* and *Htn*. At present, only *Htn* shows universal resistance.

University of Zimbabwe maize research collaboration with CIMMYT and IITA has led to the development of tropical mid-altitude, white-grain populations and inbred lines that have resistance to both northern leaf blight and maize streak virus (see below).

Maize scientists in Brazil, Egypt, India, and Pakistan, and at CIMMYT have made progress in developing subtropical germplasm that has resistance to northern leaf blight. In addition, CIMMYT's subtropical Populations 34 (Blanco Subtropical), 42 (ETO-Illinois), and 44 (AED-Tuxpeño) show good levels of resistance to northern leaf blight. CIMMYT has developed two new subtropical populations that have resistance to northern leaf blight: Population 500 is a intermediate white dent and Population 600 is a late white dent derived from various CIMMYT populations and commercial hybrids available in Mexico. A synthetic variety with a high level of resistance to northern leaf blight has been developed from each of these populations.

University of Hawaii MIR (Maize Inbred Resistance) trials conducted in various tropical locations, have led to the identification of some temperate inbred lines that have race non-specific resistance (Kim 1990).

Stalk and Ear Rots

Humid tropical and subtropical environments and high year-round temperatures favor the development of the fungal and bacterial organisms that cause stalk rots and ear rots in maize. The cold winters of temperate environments greatly diminish the presence of stalk- and ear-rotting diseases. In tropical and subtropical environments, stalk rot incidence of 5 to 40 percent is common. Early maturing germplasm is the most susceptible. Highland floury germplasm is highly susceptible to ear rot. Temperate germplasm grown in the tropics suffers severe stalk and ear rot damage.

In recent years, tropical and subtropical germplasm complexes have been improved for resistance to stalk and ear rot, and breeder have made progress in accumulating genes for resistance. However, under severe environmental stress and disease pressure, even the most resistant germplasm is affected (De Leon and Pandey 1989).

Two general types of stalk rot occur. Both are favored by warm temperatures (above 30°C) and moist soil with high nitrogen content (Renfro 1985). One type stalk rots attacks plants during the active growing stage and kill the plant by pollination time. In the very wet, warmer areas, *Pythium* spp. and *Erwinia* spp. are the important causal agents. In the intermediate and higher elevations cool and humid areas, *Diplodia maydis* is an important causal agent. Because, various species of stalk rot often simultaneously attacks the plant, polygenic resistance is required for protection. Pioneer Hi-Bred Inc. has apparently found resistance to *Pythium* stalk rot and is using it in its breeding materials in Egypt.

The second category of stalk rot is caused primarily by *Fusarium*, *Gibberella*, *Diplodia*, *Macrophomina*, and *Acremonium* species that attack the plant a few weeks before physiological maturity and develop rapidly in malnourished or senescent tissue. Generally, early maturing germplasm is more susceptible to stalk rots than late maturing germplasm. Selection for late senescence (ability to maintain green leaves until physiological maturity) has been a useful way to reduce the incidence of stalk rot. Another research approach is to select directly for genetic resistance to stalk-rot pathogens.

Most of the fungi and bacteria that cause stalk rot also infect maize ears and kernels. Although genetic variation exists for resistance (mainly polygenic), physical factors such as tight husk cover and hard vitreous grain also serve as barriers to ear rot infections. Tropical germplasm has higher levels of ear rot resistance than subtropical materials.

Good husk cover is important and can help reduce ear rots. Such control, however, is independent of the genetic resistance of an ear to infection. Also incidence of ear and stalk rot infection bears a relationship to the attack of insects such as maize earworm, borers, and fall armyworm that damage stalks and ears.

Despite the high susceptibility of temperate germplasm to tropical stalk rots, some resistance does exist, especially among inbreds developed in the southern United States and Hawaii. High resistance to *Diplodia* stalk rots is found in U.S. Corn Belt material.

A selection of the inbred line Ki3 (Kasetsart University, Thailand) derived from Suwan-1 shows ear rot resistance. Some of CIMMYT's tropical elite inbred lines such as CML 9 have good stalk strength.

Downy Mildew

Of nine species of fungi that cause downy mildew on maize, five *Perenosclerospora* species and one *Sclerophthora* species are of economic importance. The latter, which causes brown stripe downy mildew, is principally limited to India, where adequate resistance exists in released cultivars. During the 20th century, downy mildew spread to Latin America from its traditional home in Asia. It is becoming more widespread in Africa as well. Several serious outbreaks of downy mildew affected the maize crops in Thailand, India, Mexico, Venezuela, Nigeria, and the United States (Texas) during the 1960s and 1970s. Downy mildew has been especially severe in the tropical areas of Southeast Asia. In 1972, Thailand lost as much as half of its maize crop to downy mildew.

A broad range of high-yielding germplasm that has adequate resistance to downy mildew is now available for most areas where this disease is of economic importance. The resistance of new yellow-grain maize cultivars in Southeast Asia is particularly good. Maize research institutions in Thailand and the Philippines have contributed significantly to the development of such germplasm. The Thai variety Suwan-1 is perhaps the most famous downy mildew-resistant germplasm. Its resistance comes from two improved varieties from Philippines, DMR-1 and DMR-5. International testing demonstrated that low-frequency resistance existed in some of the high-yielding germplasm as well, which could be increased through a recurrent selection program for improvement of this trait. Scientists also discovered that if a plant is resistant to one *Perenosclerospora* species, there is a high probability that it will be resistant to other species. International germplasm exchange in Asia and cooperative testing for resistance to downy mildew led to rapid development and deployment of resistant varieties.

Since the 1980s, CIMMYT and the Thai national maize program have collaborated in improving the downy mildew resistance in various populations. Adequate levels of resistance have been reached in the white-grain Population 22 (Mezcla Tropical Blanca) and the yellow-grain Population 28 (Amarillo Dentado) and Population 31 (Amarillo Cristalino-2). Four new source populations (early maturing yellow and white; late-maturing yellow and white) have also been developed. Those populations were created by crossing the best performing experimental varieties adapted to the lowland tropics (CIMMYT and Thai materials) with sources of strong resistance to downy mildew, especially germplasm from the Philippines where the most virulent *Perenosclerospora* species are found. The four populations have shown good resistance to downy mildew in the Philippines and other countries of Asia.

Maize Streak Virus

Maize streak virus is a damaging disease found in Africa. It occurs in the forest and savanna zones and from sea level to 2,000 meters elevation. Streak virus is transmitted by leafhoppers (*Cicadulina* spp.), and its occurrence is closely tied to the population dynamics of the vector, which is, in turn, influenced by rainfall, temperature, and the availability of alternate hosts.

The occurrence of maize streak virus is so erratic that selection of resistant plants under natural epidemics has been ineffective because of plants that escape infection. In 1975, IITA scientists began a research program to overcome maize streak virus. They developed a methodology and facilities for large-scale leafhopper rearing and for field infestation to induce uniform streak pressure in the breeding nursery.

Streak resistance is conferred by a few major genes and thus can be incorporated in susceptible germplasm through backcrossing breeding program. IITA identified two important sources of streak resistance: IB 32, a line developed from the variety TZ-Y from IITA, and the variety La Revolution, developed at IRAT on Reunion Island. Neither of these resistant sources exhibit immune-type reactions to the disease, but they have a high degree of tolerance.

Using a backcrossing program, breeders first incorporated these sources of streak virus resistance into late-maturing, white-grain populations—TZB, composed of African and Latin American germplasm, with Nigerian Composite B being the dominant component, and TZPB, based on CIMMYT's Tuxpeño Planta Baja Cycle 7. By 1977, two streak-resistant populations—TZSR (W), a white semi-dent, and TZSR (Y), a yellow semi-flint—had been developed. Both had narrow genetic bases and poor agronomic traits. However, these populations were the progenitors of several other streak-resistant populations of different maturity periods and adaptation: TZSR-W-1, TZSR-Y-1, TZESR-W, TZESR-Y, TZMSR-W, and TZUTSR-W.

Through a collaborative research program CIMMYT and IITA are introducing streak resistance into a broad range of lowland tropical populations. Popular commercial varieties previously released by sub-Saharan national maize programs in streak-affected areas have been converted to resistant versions. In addition, tropical mid-altitude populations and inbred lines with resistance to maize streak virus have also been developed, through University of Zimbabwe and CIMMYT-IITA collaboration.

Combined Resistance to Downy Mildew and Maize Streak Virus

With the spread of the downy mildew in West Africa, IITA has been working to develop high-yielding germplasm that possesses combined resistance to downy mildew and maize streak virus. Two late-maturing white and yellow-grain populations (DMR-LSR-W and DMR-LSR-Y) and two intermediate to early white-grain (DMR-ESR-W and DMR-ESR-Y) populations are the result of this work. Varieties and lines that have downy mildew resistance derived from Thailand's population Suwan-1 have also been converted to streak-resistant types. DMR-ESR-W has been released in Nigeria, Benin, and Togo and is becoming popular among farmers. More work is still needed to develop a range of downy mildew-resistant, white-grain genotypes for West Africa.

Corn Stunt

Stunt, originally called Rio Grande corn stunt, is an important disease in tropical environments in Mexico, Central America, the Caribbean, and in the northern countries of South America. It is caused by corn stunt spiroplasma and is spread mainly by the leafhopper *Dalbulus maydis*.

Sources of germplasm with resistance to stunt exist at CIMMYT, national maize programs in Mexico, Central America and the Caribbean, and the state universities of Florida, Hawaii, and California. Some inbred lines, synthetic varieties, and hybrids have been derived from these stunt-resistant populations. The experimental variety Santa Rosa 8073, selected by the Nicaragua national program, has been especially outstanding.

Stunt-resistant germplasm has dramatically changed the varietal picture in Central America. In Nicaragua, the first stunt-resistant open-pollinated variety, NB-6, was released in 1984, and was followed by NB-12 in 1987. By 1991, 80 percent of Nicaraguan maize farmers were planting stunt-resistant varieties. Today, the spread of stunt-resistant germplasm throughout Central America and the Caribbean has virtually eliminated the disease as an economic threat.

Late Wilt

Late wilt is found mainly in Egypt and India. During the 1950s, late wilt had a devastating impact on the maize crop in Egypt because the prevailing double-cross hybrids then in commercial use were highly susceptible. Egyptian maize scientists soon developed a number of elite populations, open-pollinated varieties, and inbred lines with resistance. Resistant germplasm is derived from Egypt's American Early Dent (AED) population, which is based on the U.S. dent variety Boone County White. This material was combined with CIMMYT's Population 21 (Tuxpeño) to create Population 44 (AED-Tuxpeño), which has produced some high-yielding varieties and inbred lines for subtropical environments.

Insect Resistance

Insects generally are more damaging in farmers' fields in the tropics and subtropics than diseases. CIMMYT data on the maize-producing environments of the developing world indicate that in 29 largest maize-producing countries, over half of the area under maize is seriously affected by insect problems.

Despite the existence of genetic resistance, progress in breeding for host plant resistance to economically important insects has been slower than breeding for disease resistance. The lag has been partly due to the need to develop methodologies for rearing insects and to find cost-effective ways to apply larvae to plants to ensure uniform infestation. Recently, diets, equipment, and procedures have been developed that allow efficient uniform insect infestations and the screening of thousands of plants (Mihm 1989). The recurrent breeding methodology and artificial infestations are being used by several maize research programs to improve levels of host plant resistance to important maize pests. The emphasis is on improving insect resistance levels with good agronomic performance and stability of performance.

Corn Borers

Genetic variability exists within the maize genome for borer resistance. Pioneering work was done at Mississippi State University, using Antigua and southern U.S. germplasm. CIMMYT studies support the origin of a generalized resistance to borers in Antigua germplasm and suggest that important chromosomal regions controlling this resistance are located on chromosomes 1 (L, long arm), 2, 3 (L), 5 (L), 10 (L), and 9 (S, short arm). These studies further indicate that the resistance to borers is polygenic and primarily additive in gene action and that there are large genotype x environment interactions.

First-generation resistance to southwestern corn borer (*Diatrea grandiosella*) appears to impart a level of resistance to other borers, such as *Ostrinia nubilalis*, *O. furnacalis*, *D. saccharalis*, *Chilo* spp., *Busseola* spp. and *Sesamia* spp. as well as to fall armyworm (*Spodoptera frugiperda*). Results so far indicate moderate to highly stable resistance to individual and multiple species of borers. Resistant varieties show significantly less leaf feeding damage than nonresistant varieties. Two notable CIMMYT borer-resistant sources are the subtropical Multiple Borer Resistant (MBR) Population 590 and the Multiple Insect Resistance Tropical (MIRT) Population 390.

The IITA-developed populations TZBR-Sesamia-1 and TZBR Sesamia-3 are proving to be good sources of resistance to the stem borer *Sesamia calamistis* (Mareck, Bosque-Perez, and Alam 1989). Two other populations, TZBR-Eldana-1 and TZBR-Eldana-2, are the best sources of resistance to *Eldana saccharina* (Bosque-Perez and Mareck 1990). Both of these borer species are restricted to Africa.

Technology for the application of DNA markers has the potential to improve selection efficiency and accelerate the rate of progress of germplasm with good agronomic performance and adequate levels of host plant resistance. RFLP markers are being used to identify chromosomal regions that control quantitative traits loci that affect insect resistance. A cost-saving and gain in efficiency is expected if RFLP markers can be used in practical breeding to facilitate the incorporation of insect resistance into elite germplasm.

In temperate zones, progress has been achieved in breeding for resistance to the first and second generations of the European corn borer* (Gracen 1986). Some progress has also been made in breeding for resistance to southwestern corn borer. Scientists at Iowa State University and the USDA research center in Ankeny, Iowa, have developed several synthetics, referred to as the Iowa Corn Borer Synthetics, that have resistance to both generations of the European corn borer (ECB). Scientists in New York have developed Cornell ECB Composite, which has shown high levels of resistance to first and second generation European corn borer damage. This population was developed by crossing insect-resistant germplasm from CIMMYT (largely Antigua and Caribbean germplasm) with U.S. single-cross hybrids.

Temperate lines exhibiting resistance or tolerance to Oriental corn borer are NC248 (North Carolina), Hi32 (Ohio/Hawaii), Hix4231 (Hawaii), and Mo5 (Missouri) (Kim 1990).

Fall Armyworm

In the Western Hemisphere, fall armyworm can be one of the most destructive maize field pests in tropical and subtropical environments and also in temperate zones in certain years. CIMMYT has been working with several national maize programs to develop fall armyworm resistance for more than a decade. Progress has been slow. Recently, much heavier selection pressure for fall armyworm resistance is being placed on special-purpose populations.

Corn Earworms

Several species of corn earworms exist, principally *Heliothis zea* and *H. armigera*. This insect is widespread, but poses a serious problem only for floury maizes grown in high valleys of the Andes. Sources of resistance to corn earworms have been found in the Mexican races Zapalote Chico and Zapalote Grande. Both materials are low-yielding, partly because of their susceptibility to common rust and northern leaf blight. The race Zapalote Grande is the most promising of the resistant germplasm sources identified thus far.

---

* Table 3.5 provides Latin names of major maize insects.

Stored-grain Pests

Grain weevils can cause significant economic damage to maize grain, especially in the humid tropics and subtropics. Dent grain genotypes are more vulnerable than flints. Dent grain genotypes have softer endosperms, which make it easier for a grain weevil to lay eggs in the grain and for the larvae to damage the kernel. Tight, undamaged husk cover also reduces weevil infestation in the field. Unfortunately, many of the highest-yielding improved tropical maize materials, e.g., the Mexican dents, have relatively poor husk cover, which makes them highly susceptible to weevil damage.

Differences have been found among genotypes for resistance to stored-grain insects, but chemical rather than host-plant resistance has been the primary method used for controlling such pests. The factors in genetic resistance to grain weevils are grain hardness, sugar content, and the presence of phenolic compounds in the aleurone layer. CIMMYT, in collaboration with Canadian researchers, has identified 15 materials, 10 of which are quality protein maize, that have intermediate levels of resistance (Mihm 1990).

IITA is collaborating with the UK Natural Resources Institute to identify sources of resistance to the maize weevil and the larger grain borer. The larger grain borer has been a mounting problem in West Africa since it was introduced from Central America in the 1980s.

Resistance to Striga

Striga, a parasitic weed, is a costly pest of maize production, particularly in sub-Saharan Africa. Five species of striga infest maize. *Striga hermonthica* and *S. apera* are the most important. Striga does the greatest damage to maize in low-fertility soils. Until recently, crop rotations and fallow were the main control methods. Progress in resistance breeding was hampered by the problem of artificial infestation. IITA has now developed infestation techniques that allow the exertion of heavy and uniform pressure on materials being screened. The best sources for striga resistance thus far are found mainly in U.S. germplasm. Most tropical dent varieties are susceptible, although some segregation for resistance has been observed. All Tuxpeño groups are highly susceptible. Some IITA hybrids show moderate-to-good levels of resistance.

Drought Tolerance

About 95 percent of the maize area in the tropics is dependent on rainfall, which tends to vary considerably from season to season. Moisture stress is widely experienced by maize grown in these tropical rain-fed environments. Although the average rainfall in a region may appear adequate, dry spells can occur at any time during growing season.

Maize is most vulnerable to moisture stress that occurs in the period lasting from 2 weeks before to 2 weeks after flowering. During this period, drought depresses yield potential by irreversibly limiting the number of kernels and ears that develop. Moisture stress during flowering lengthens the interval between anthesis and silking and decreases the number of silks that are viable for pollen germination to fertilize the embryos.

In effect, as a result of moisture stress during flowering, the plant produces fewer kernels than it would have if moisture and other production conditions had been more favorable. The plant aborts ears and grain and concentrates its limited energy to assure male flowering and pollen shed, increasing the odds that some pollen will find embryos to fertilize in surrounding plants that are not so moisture-stressed. Under moisture stress, it is not uncommon for the ear to sacrifice the development of more than half of its potential kernels. In severe cases there may be no seed set at all.

Finding ways to ensure greater seed set—by reducing the anthesis-silking interval—and survival in drought-stressed maize plants is central to the development of drought-tolerant cultivars (Bolaños and Edmeades 1993). Fortunately, sufficient genetic variation exists in maize germplasm to permit selection of progenies that are more capable of synchronizing male and female flowering under stress, and thus have increased grain production.

A network of national programs (India, Brazil, China, Ghana, Burkina Faso, Côte d'Ivoire) and CIMMYT and IITA are collaborating to develop elite maize populations with enhanced drought tolerance. The objective is to produce germplasm that has improved seed set under moisture stress but retains high yield potential under more favorable growing conditions.

Among the seven elite populations CIMMYT is improving for drought tolerance, the most outstanding is Tuxpeño Sequia. Over eight cycles of improvement, the grain yield of Tuxpeño Sequia under medium moisture stress rose 20 percent and under severe moisture stress it rose 39 percent (Table 7.2). However, the yield performance of Tuxpeño Sequia under moisture stress after cycle 6 has changed little, suggesting that genetic variability for drought-tolerant traits is becoming exhausted in the population. By comparison, IPTT 21, a population selected under nonstressed conditions, showed essentially no change in yield under moisture stress over six cycles of selection.

TABLE 7.2 Changes in grain yield during six to eight cycles of selection in Tuxpeño Sequia and IPTT 21 under mild, medium, and severe soil moisture stress (Tlaltizapan, Mexico).

	Yield (t/ha)		
Selection cycle	Mild stress	Medium stress	Severe stress
Potential yield level	6.0	4.0	2.0
	Tuxpeño Sequia		
C_0	5.27	3.83	1.67
C_4	5.89	3.95	2.02
C_8	6.84	4.58	2.32
	IPTT 21 [a]		
C_0	5.27	3.83	1.67
C_6	5.31	3.53	1.75

[a] Population 21, International Progeny Testing Trial.
Source: Bolaños and Edmeades 1993.

IITA has screened the world collection of tropical inbred lines for drought tolerance and formed a drought-tolerant synthetic from 12 selected drought-tolerant lines. Among the inbred lines that were tested, the best-performing were KU 1414 from Thailand and Tzi 9 from IITA. The yield advantage of the drought-tolerant synthetic was apparent in multilocation testing under soil moisture stress in West Africa (S. K. Kim, personal communication).

Cold Tolerance

Maize is grown under cool temperatures in parts of developing world such as the high-elevation highlands of Mexico, Ecuador, Peru and Bolivia; the Himalayas stretching from Afghanistan to China; and the winter season in India, Pakistan, Vietnam, and southern China. In these environments, average day-time temperatures can drop below 15°C and minimum night-time temperatures occasionally fall below zero, causing frost. Prolonged low temperatures arrest photosynthesis, resulting in pale green and yellow leaves and chlorosis.

The effect of low temperatures on plant development depends on the stage of growth at which the plant is exposed and the severity and duration of the stresses. Research in the United States and Pakistan has shown that genotypes that have cold tolerance also exhibit tolerance to higher than normal temperatures.

Maize scientists in Mexico, Pakistan, India, and New Zealand are working to develop high-yielding germplasm with improved cold tolerance, both for highland environments and for winter-season maize grown in subtropical environments. Good cold tolerance can be found in some temperate maize genotypes, and some highland genotypes exhibit a degree of tolerance to near-freezing temperatures and frost.

Highland Maize

To survive cool day and night temperatures, highland maize has developed adaptive mechanisms, including an efficient enzyme system that allows the plant to continue metabolizing at temperatures too low for most other maizes.

Several highland gene pools of Mexican or Andean origin have shown themselves to be promising sources of cold tolerance. National maize programs in Latin America (notably Mexico and Ecuador) and CIMMYT are further improving highland germplasm for cold tolerance, using deep purple stem color as the primary visual selection criterion because dark stem colors absorbs a higher energy part of the sunlight spectrum and increases plant temperatures. From this work, highland Population 900 shows excellent cold tolerance.

Winter Maize

Increasing numbers of farmers in India, Pakistan, Bangladesh, Vietnam, and southern China, are planting a crop of maize is in the cool, dry, winter season. Maize planted in December-January usually suffers from cold stress, which inhibits seedling emergence and stand establishment and early plant growth. Later in the season—April or May—the temperatures rise rapidly, particularly during the day, and even though most winter maize is grown under irrigation, the plants suffer moisture stress. Farmers need drought-tolerant winter maize varieties to reduce the total amount of irrigation water required and thus reduce the cost of cultivation.

In India, maize breeders at Punjab Agricultural University (PAU) at Ludhiana developed the population, Partap, from crosses among three hybrids: Vijay x Pioneer 102, Pioneer 114 x Vijay, and Ganga 5 x Pioneer 104 (Khehra et al. 1988). This population showed good variability for grain yield and cold tolerance. The variety Partap (C4), which consistently yielded 15 percent higher than check Composite Vijay, was released in 1988 for use in the Indo-Gangetic plains. PAU breeders have established two heterotic pools, Ludhiana Lancaster Pool and the Ludhiana Stiff Stalk Pool, for use in developing cold-tolerant inbred lines (Khehra et al. 1988).

In Pakistan, national maize breeders are using temperate germplasm from the United States and CIMMYT subtropical Population 45 to develop cold-tolerant germplasm with traits for tolerance to other stresses.

Aluminum Tolerance

Tropical maize, mostly in South America, is planted on at least 6 million hectares of acidic soils that are heavily leached of nutrients, have a low base exchange capacity, and are low in available phosphorus. Some acidic soils also contain amounts of soluble or exchangeable aluminum that are toxic to maize plants (toxicity depends on concentrations of Al, P, Ca, and Mg and soil temperature). Excessive aluminum in the soil severely limits the capacity of roots to absorb essential nutrients and moisture by interfering with the plant's biological and physiological processes.

Deep liming the soil to about 30 centimeters can raise soil pH and reduce aluminum toxicity by bringing about aluminum precipitation. But the cost and logistical problems of obtaining lime, especially for resource-poor farmers, and the incompatibility of deep liming with conservation tillage practices, have prompted maize breeders to seek variation in the maize genome for aluminum tolerance.

Brazilian researchers from the National Maize and Sorghum Research Center (CNPMS) have shown the presence of aluminum tolerance in a range of germplasm. They have developed a broad-based population, Composto Amplo, that has produced some outstanding Al-tolerant germplasm, such as BR 451, which has become one of the most popular hybrids grown in the Cerrado.

CIMMYT scientists stationed in the South American regional maize program, headquartered at Cali, Colombia, have developed seven Al-tolerant populations, which are being improved for resistance to rusts and northern leaf blight. Some of the varieties and hybrids developed from these materials have yielded up to 200 percent better than nontolerant materials in soils with medium to high levels of aluminum. Among the yellow grain germplasm, Populations SA-4 and SA-5 form a heterotic pair. Among the white grain germplasm, Populations SA-6 and SA-7 form a heterotic pair.

The IITA/Cameroon maize project, supported by USAID, has formed an acid-tolerant population, and selection has been done in acid soils in the country's western highlands (L. Everett, personal communication).

Early Maturing Tropical Germplasm

Early maturing genotypes account for about 30 percent of the lowland tropical maize area. Varieties with 90- to 100-day maturity are planted on 8.4 million hectares in the lowland tropics and varieties with 80- to 90-day maturity are planted on another 2.5 million hectares. Approximately 50 percent of this area is found in Asia, and the rest is evenly divided between Africa, and Latin America.

The demand for early maturing varieties is driven by both economic and biological forces. Early maturing varieties allow farmers to fit maize into more intensive cropping patterns, e.g., patterns in which two or three crops are grown annually, they provide a form of drought escape in areas where the rainfall period is too brief for long-season varieties, and they give farmers an early source of food after the dry season.

Primitive landraces include some very early maturing maizes. But all are low yielding, tend to be susceptible to disease, and they suffer more insect damage (due to earlier plant development than surrounding plant types). The great susceptibility of early maturing germplasm to biotic stresses slows the development of improved germplasm that yields well and makes it difficult to extract vigorous inbred lines for hybrid development.

The challenge for breeders has been to develop early maturing varieties that combine high yield potential with disease and insect resistance. The most successful approach has been to assemble a number of high-yielding intermediate-to-late maturity genotypes into a population, and then conduct a recurrent selection program with earliness as the major criterion used to reconstitute the next cycle.

Several national maize programs in Asia and Africa are working with CIMMYT and IITA to develop germplasm that combines early maturity and high yield potential. The best early maturity germplasm complexes thus far developed are more correctly termed intermediate-to-early maturity materials. These elite populations have a maturity period of 90 to 100 days, a yield potential of 4 to 5 t/ha, and resistance to several important diseases.

Improved Nutritional Quality

The discovery in 1963 of the recessive mutant gene, opaque-2, which doubles the levels of lysine and tryptophan in maize endosperm excited scientific circles. Raising the levels of these two essential amino acids, which normally limit metabolism of maize protein, promised to significantly improve the nutrition on humans and monogastric animals that consume maize-based diets. The opaque-2 gene, however, brought along some unwanted characteristics as well. Opaque-2 maize was low-yielding and had soft, chalky-dull grain. It was susceptible to ear rots and stored-grain insects and slow to dry down after physiological maturity. These negative characteristics discouraged farmers from adopting opaque-2 genotypes.

TABLE 7.3 CIMMYT advanced quality protein
maize populations for tropical and subtropical
environments.

Name	Description
Tropical adaptation	
61 Early Yellow Flint QPM	EYF
62 White Flint QPM	LWF
63 Blanco Dentado-1 QPM	LWD
64 Blanco Dentado-2 QPM	LWD
65 Yellow Flint QPM	LYF
66 Yellow Dent QPM	LYD
Subtropical adaptation	
67 Temp. Blanco Cristalino QPM	LWD
68 Temp. Blanco Dentado QPM	IWD
69 Templado Amarillo QPM	IYD
70 Templado Amarillo Dentado QPM	IYD

Key: L = late maturity. F = Flint. I = intermediate maturity.
D = dent. E = early maturity. W = white. QPM = Quality
Protein Maize. Y = yellow.
Source: CIMMYT.

In the 1970s, CIMMYT scientists formulated procedures to develop normal-looking, normal-tasting quality protein maize (QPM) genotypes that retain the nutritional quality of the original opaque-2 maize. They employed a painstaking modified backcrossing program to accumulate of a myriad of quantitatively inherited modifier genes and were able to overcome most of the deleterious linkages between the opaque-2 gene and various yield and grain quality characteristics. By the 1980s, a sizable collection of tropical and subtropical QPM gene pools and elite populations had been developed that had high yield potential, hard endosperm vitreous grain, and much improved resistance to diseases and insects. In the late 1980s, a range of QPM inbred lines were developed for hybrid combinations.

Although CIMMYT's QPM maize program was suspended in 1991, pools, populations, and inbred lines previously developed (Table 7.3) are being supplied to national programs and private companies interested in such research. Brazil, China, South Africa, and Ghana are among the national maize programs that are still vigorously pursuing QPM research. About 500,000 hectares are planted to QPM material.

8

Building Successful Maize Seed Industries

About 60 percent of the world's maize area is planted to high-yielding hybrids (CIMMYT 1994), which are the greatest practical achievement of plant breeding research to date. Another 10 percent of the area, entirely in developing countries, is planted to improved open-pollinated varieties. (The remainder is planted to traditional varieties.) In developed countries, nearly 100 percent of the maize is hybrid. By contrast, in the developing countries, hybrids are planted on about 44 percent of the maize area. Argentina, Brazil, and China account for nearly four-fifths of that. Of the 54 developing countries that grow over 100,000 hectares of maize, only in 15 are improved genotypes regularly planted on over half the maize area (Table 8.1).

The regions in the developing world with the heaviest use of commercial seed of improved maize genotypes are East Asia and the Southern Cone of South America. Both have relatively well-developed agricultural systems and mainly temperate maize production environments. The lowest rates of improved seed use occur in western and central Africa and Andean South America. In these regions, maize is primarily grown under low-fertility conditions by resource-poor farmers for their own consumption.

Only in a few developing countries (Thailand, Egypt, Guatemala, India, Nigeria) have seed industries been successful in marketing improved open-pollinated composite and synthetic varieties. However, even in these countries, most farmers have switched to nonconventional and conventional hybrids as they have become available. Still, improved open-pollinated varieties retain a market niche, especially among resource-poor, small-scale farmers.

In most developing countries where hybrids are grown, double crosses and three-way crosses are the predominant genotypes in use. (In the industrialized countries, single-cross hybrids replaced double and three-way crosses two decades ago.) But in most developing countries, the difficult growing conditions and low seed production (and even survival) of weak inbred lines have made single-cross hybrid seed production less profitable and practical.

Even though improved open-pollinated varieties and hybrids have been developed for the major maize production environments in the Third World, discriminatory seed price policies and the presence of public seed enterprises have precluded the entry of private companies into maize seed production and marketing in the majority of developing countries. Without more effective and dynamic maize seed industries, progress in the adoption of other crop management innovations, such as fertilizers, pesticides, and farm machinery will be stymied and few productivity gains—or agricultural development—will occur.

Importance of Improved Seed

Maize researchers have developed elite open-pollinated varieties and hybrids with continually rising yields, improved disease and insect resistance, and stronger stalks. These traits are carried from one generation to the next in the seed. Until farmers harvest a superior crop from the seed of improved varieties and hybrids, no one benefits from the

TABLE 8.1 Important maize-growing developing countries where over half of the maize area was planted to improved open-pollinated varieties (OPVs) and hybrids, 1992.

Country	Area (%) planted to improved commercial		
	OPVs	Hybrids	Total
Asia			
China	7	90	97
India	49	14	63
Thailand	80	15	95
North Korea	–	95	95
Vietnam	50	10	60
Latin America			
Brazil	13	44	52
Argentina	8	85	93
Venezuela	5	95	100
Chile	5	85	90
Africa			
Mozambique	61	4	65
Senegal	98	0	98
Kenya	11	74	85
Zimbabwe	0	100	100
Egypt	50	28	78
Zambia	5	65	70

Source: CIMMYT 1994.

time and money invested in developing an improved genotype. But unless seed of improved, management-responsive varieties is available, farmers have little incentive to spend money on fertilizer and other productivity-enhancing cultural practices. Thus, in the absence of a maize seed industry capable of producing and delivering quality seed of appropriate genotypes to farmers in an efficient and dependable manner, the payoff from investments in maize breeding research can not be realized.

The characteristics of hybrid varieties make the existence of an organized seed industry essential for their diffusion. The quantum yield jump (heterosis) achieved by crossing genotypes that have good combining ability is expressed only in first generation (F_1) hybrid progenies. If the F_2 seed and subsequent seed generations are planted, the yield may drop by 30 percent. Thus seed production and marketing must be proven reliable enough to assure farmers that F_1 seed of hybrids will be available every year. Also the controlled cross pollination necessary for hybrid seed production requires a skilled work force to detassel plants. Such work is facilitated by an organized seed industry.

When farmers plant their own seed from open-pollinated varieties or certain classes of nonconventional hybrids, the yield depression that occurs is hardly noticeable. But in farmers' fields because of pollen contamination from other nearby varieties, even the seed of an open-pollinated maize variety should be replaced every 3 years or so to maintain the genetic superiority inherent in the original parent variety. In contrast in self-pollinating crops like wheat and rice, each succeeding progeny is genetically identical to the parent (assuming that the parent is a homozygous line). Thus, year after year, the farmer can save the seed of one crop to plant the next crop, unless it succumbs to a new disease or insect pest.

Seed Types

Different maize seed types are grown by farmers around the world, particularly in developing countries. Although specific names are used to described these seed types, all are maize varieties in the generic sense of being individual plants or groups of plants that are distinct from other groups and identifiable from generation to generation.

A traditional variety is an open-pollinated seed type that has been selected and maintained by the farmer. At harvest, the farmer selects ears from the best plants, based on criteria such as grain yield, color, freedom from disease and insect damage, size of ear, or forage yield. Over generation under this system of farmer care and management, maize became radically altered in appearance and many other morphological characteristics. A traditional variety usually attains its highest yield potential in the environment where it was selected and often has considerable genetic buffering capacity to withstand the stresses peculiar to that location. But, a traditional variety tends to have a limited breadth of adaptation. In other words, its yield potential drops precipitously when it is grown in environments that differ slightly from its original home. Part of the reason for this is the narrow genetic base of many traditional varieties.

Improved Open-pollinated Varieties

Sophisticated selection procedures have permitted breeders to develop superior new maize populations (germplasm complexes) through controlled (or directed) crossing of a large number of breeding lines and populations that in theory have good combining quality and desirable genetic characteristics. Subsequent population improvement by recurrent selection methods further concentrates the desired genes.

Today, breeders develop an improved open-pollinated variety (OPV) from a recombined fraction of relatively uniform phenotypes that represent the superior fraction of an improved maize population. In this technique, the OPV will be constituted by recombining 8 to 10 selected families (all plants in each family come from the same ear) that have similar maturity, plant height, ear height, and other characteristics. They can produce plants that have a quite uniform appearance. Even though the elite fraction of the population is recombined to produce the OPV, it is recognized that the variety will show some variation for important agronomic characters.

Another type of open-pollinated variety is called a synthetic variety. This OPV is derived by combining several early generation self-pollinated inbred lines (S_2 or S_3), selected for their combining ability. These inbred lines produce a new high-yielding population that subsequently is improved through some form of recurrent selection.

The development of improved open-pollinated maize varieties, has helped to compensate for the weak maize seed industries in many developing nations. Improved open-pollinated varieties can be widely distributed, even from farmer to farmer, without suffering losses in yield potential or agronomic desirability. By outcrossing with traditional varieties, the improved open-pollinated varieties have helped to introgress superior germplasm—and higher yield potential—into farmer-maintained varieties in virtually every maize-growing area in the developing world.

Seed multiplication plots for open-pollinated varieties are grown under high-yielding crop management practices. They can be planted like a normal field of maize, although they must be sufficiently isolated (up to 200 meters from other varieties) to prevent pollen contamination. The yields from seed plots of high-yielding open-pollinated varieties—after roguing, discards, and seed cleaning—are within 85 percent of the yield of a commercial field of maize.

Hybrids

Hybrids fall into two broad categories: conventional and non-conventional. Conventional hybrids result from crossing two or more inbred lines that have high combining ability. Inbred lines are developed through a process of self-pollination, called inbreeding, which usually is repeated for four or five successive generations. The more inbreeding a line has undergone, the shorter, weaker, and lower yielding it becomes.

Single-cross hybrids are the highest yielding of all the maize seed types. But inbred lines—the parents of the hybrid—usually are weak and when crossed the resulting seed yield is very low. These shortcomings kept single-cross hybrids from being a commercial practicality for more than 30 years. It was not until maize breeders developed more vigorous germplasm complexes capable of producing vigorous inbred lines with higher yield potential that single-cross hybrids became an economic reality. The seed yield of single-cross hybrids is still low (about 50% of OPV seed yield) so it is important to multiply inbred lines under the best possible production conditions.

The first hybrid to achieve commercial success was a double-cross hybrid. Invented early in the 20th century, the double-cross involves the union of two single-cross hybrids, themselves each the product of two inbred lines (Fig. 8.1). The double-cross hybrid solved the problem of the low seed production inherent in inbred-line seed multiplication, but it added another step in the seed production process. The *seed* yield of a double-cross hybrid is 2.5-times that of a single-cross hybrid (and 25% more than OPV seed yield). Because double-cross hybrids involve four inbred lines and an additional cycle to produce certified seed, they have tended to be replaced by three-way cross hybrids, which are less complicated to produce.

A three-way cross hybrid involves the union of an inbred line, serving as the male, and a highly productive (in seed yield) single-cross hybrid, serving as female. The seed yield when a three-way cross is produced (10% more than OPV seed yield) may not always be as high as the seed yield when double-cross is produced. But when a three-way cross hybrid seed is planted, it yields somewhat more than double cross hybrids and the plants are more uniform.

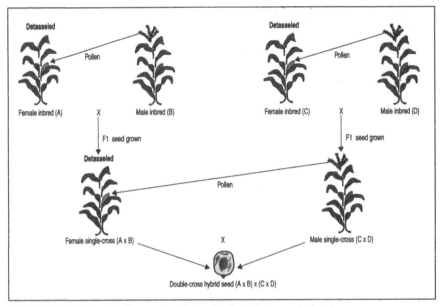

FIGURE 8.1 Steps in the production of a double-cross hybrid.

Producing hybrid seed is considerably more complicated and costly than producing seed of open-pollinated varieties. Special planting patterns (4 female:1 male for a good pollen-shedding parent and 3 female:1 male for a poor pollen-shedding parent) are required. The controlled pollination phase must be carefully managed and closely supervised.

Various nonconventional hybrids are also grown in developing countries. These hybrids are formed through crosses in which at least one parent is not an inbred line. A variety-cross hybrid is a cross between two open-pollinated varieties that are sufficiently different genetically to show hybrid vigor. A family-cross hybrid is formed through the union of two elite fractions that have been extracted from improved open-pollinated populations and varieties and that have high combining ability. A top-cross hybrid is a cross between a variety (generally used as the female) and an inbred line (serving as the male parent).

Nonconventional hybrids take less time to develop than conventional hybrids. Seed production is also simpler, less risky, and lower in cost per unit than the production of conventional hybrid seed. Nonconventional hybrids are less uniform than conventional hybrids and have a 10 to 15 percent lower yield potential. Nonconventional hybrids are usually planted in a row pattern of 1 male:4 females. Of course, all female rows must be detasseled. The relatively vigorous female seed parents result in high seed yields in these nonconventional hybrids (85 to 95% of OPV seed yield), which is another factor that makes them attractive in emerging national seed industries.

Choosing the Right Seed Type

Farmers are generally willing to adopt new cultivars when they offer tangible benefits and seed is reasonably priced. Whether a farmer adopts a new open-pollinated variety or hybrid or decides to continue to plant a traditional variety is primarily an economic choice conditioned by additional factors such as the availability of quality seed, cost of the seed and complementary inputs (mainly fertilizer), expected yield and crop profitability, ability to finance purchase of needed inputs, and the risk of crop loss during the growing season.

Serving Different Farmer Groups

In most developing countries, both open-pollinated varieties and hybrids (conventional and nonconventional) have useful roles in meeting the different production circumstances of farmers. Those roles will change as progress in research occurs, as the national seed industry matures, and as the economic circumstances of maize farmers improve. But even as more farmers buy commercial seed, and switch to hybrids, demand for open-pollinated varieties may continue. For example, open-pollinated varieties may be more cost-effective than hybrids in areas where the crop is grown both as a green fodder and grain crop. Or farmers who plant maize at high planting densities and then progressively thin the stand to obtain green fodder for livestock, a frequent practice in the Himalayan region, may find hybrid seed too expensive, given the end uses of the crop. For such farmers an open-pollinated variety is preferable because the seed can be retained from the previous crop and, when seed has to be purchased, it is less costly.

To reach small-scale farmers who remain outside commercial seed channels, artisan seed production and distribution schemes are appropriate. To establish and support artisan maize seed production programs, research and extension services can help by recruiting farmer-seedsmen and training them in seed selection and preservation procedures and by supplying basic seed. Artisan seed production systems are especially appropriate in the first two stages of seed industry development (see Chapter 5) and to serve resource-poor farmers in low-yielding production environments and isolated areas.

Seed Cost

Seed is not a major cost in a farmer's maize production budget. It usually accounts for 3 to 7 percent of the variable costs of inputs. As shown in Table 8.2, the farmer pays about US$3/ha for seed of his traditional variety compared with $10/ha for seed of an improved open-pollinated variety. But the farmer can save and replant seed of the open-pollinated variety for several years (at about the same cost as the traditional variety if he selects the seed stock carefully). A single-cross hybrid, at $38/ha, is certainly too expensive for a farmer who expects yields of only 2 t/ha, but not for a farmer who expects over 5 t/ha. Access to improved seed, rather than price, is often a more significant determinant of the seed type best suited to an individual farmer.

Although the yields of open-pollinated varieties and hybrids vary considerably depending on environmental conditions and management practices, the various seed types based on the same genetic stock—from open-pollinated varieties through the various classes of hybrids—show a corresponding upward progression in grain yield potential, especially when the seed is grown under good growing conditions.

The yield gap between hybrids and open-pollinated varieties narrows considerably when growing conditions become more harsh. Under the poorer crop management and unfavorable production conditions found in many tropical environments—where average yields are less than 1.5 t/ha—hybrids are rarely profitable to use, even if they show some yield superiority over improved open-pollinated varieties.

Open-pollinated varieties have a distinct advantage where seed industries are in their infancy and seed distribution is difficult and costly. Seed production costs for improved open-pollinated varieties are rela-

TABLE 8.2 Grain yield potential of different seed types relative to yield of improved open-pollinated varieties (OPVs) and average seed costs in developing countries, 1992.

Seed type	Yield relative to yield of improved OPVs (%)	Seed cost[a] ($/ha)
Farmer-maintained varieties	70-80	3
Improved OPVs	100	10
Varietal or family hybrids	105-110	12
Top-cross hybrids	110-115	16
Double-cross hybrids	115-120	20
Three-way cross hybrids	120-125	29
Single-cross hybrids	125-130	38

[a] 20 kg/ha seeding rate.
Source: CIMMYT 1994.

tively low and seed stocks can be built up rapidly. Further, because OPV seed can be replanted without suffering significant yield depression, it can be disseminated through farmer-to-farmer distribution.

Hybrids have a clear advantage in better production areas. Because hybrid seed must be replaced every year, adoption by farmers helps to develop a national maize seed industry capable of financing an expanding research agenda and of providing farmers with a continuing stream of more productive and dependable maize seed types.

Seed Supply Systems

Seed systems involve numerous organizations and specialists (Douglas 1980). The following is an overview of the activities and functions involved in the development and delivery of seed of improved maize genotypes to farmers.

Maize Breeding Research

A dynamic maize research program is the foundation of any successful seed program. Maize breeding research is a long-term activity supported by the combined efforts of plant breeders, geneticists, plant pathologists, entomologists, agronomists, and economists. The plant breeder's first job is to identify the type of maize varieties farmers require. Farmers' varietal needs are characterized in terms of grain color and type, maturity period, and resistance or tolerance to diseases, insects, and various abiotic stresses (drought, soil toxicities). The next major activity is to assemble superior germplasm complexes containing the genetic variability needed to achieve stated breeding objectives. Improved maize germplasm, varieties, inbred lines, and hybrids have been developed by publicly funded national and international and national maize breeding programs as well as by private transnational and national seed companies.

Varietal Evaluation and Release

For farmers to be interested in adopting improved varieties or hybrids, their yield potential and other traits must consistently be superior to those of farmers' current varieties. Therefore, it is vital for maize breeders to take part in the testing and evaluation of elite experimental varieties in farmers' fields and in the process of deciding whether a new variety or hybrid should be recommended for formal release to farmers.

Variety evaluation and release systems vary from country to country. In most developing countries, the growing conditions (soil, water availability, machinery) at experiment stations differ so greatly from farm conditions that any assessment of variety performance without off-station trials may be unreliable. The initial stages of varietal testing usually are the responsibility of the plant breeder who plants a large number of breeding materials at a few locations.

An effective variety evaluation system should facilitate the identification of superior varieties as rapidly as possible. Multilocational test data can be used as a substitute for testing the same variety or hybrid over a number of years. Simple on-farm variety trials can also form the basis for agronomic recommendations to be used with a particular variety.

Where several public and private organizations are involved in maize breeding research, a separate agency often is responsible for the final testing of materials developed by research organizations to help ensure that such evaluations are uniform and unbiased. For these evaluations, it is important to identify the environments for which individual varieties are suited and to test varieties with similar adaptation and maturity characteristics together.

Most developing countries have laws or regulations regarding the registration and release of new maize varieties and hybrids for commercial use. Varietal registration and release regulations are supposed to protect the plant breeder and farmer against inferior or fraudulent genotypes. A variety review and release committee establishes guidelines for adding new varieties and hybrids (and the discontinuation of old obsolete or undesirable ones) developed by public and private maize breeding programs. Results of field testing are used by variety review and release committees to assess the performance of nominated varieties, determine their potential contribution to national agriculture, and make recommendations regarding their release and entry in the seed multiplication and production scheme. The overriding principle in seed certification programs should be to facilitate the delivery of superior materials to farmers as quickly as possible.

The breeder is responsible for properly describing all important characteristics of a new variety. This information is used to monitor the genetic purity and stability of subsequent generations during breeder's seed production, Seed certifying and enforcement organizations use the information to assess the purity of the variety and to determine whether the seed is properly labeled. Plant parts that may be considered in the description of a variety include the stem, leaves, tassel, ear, and seed. Attributes such as plant color, tassel color, tassel size and configuration, seed shape, leaf orientation, stem pigmentation, ear shape, silk color, and midrib color are among the characteristics most commonly used in varietal descriptions.

Varietal Maintenance and Initial Seed Increases

The maize breeder should be responsible for maintaining the parental material of an improved open-pollinated variety or inbred line and for breeder's seed production—the initial seed-increase step in certified seed production program—for as long as the variety or inbred line remains in commercial production. When a released variety or inbred line is replaced by a superior one, seed maintenance can be discontinued. Parental material refers to the seed from a limited number of maize plants used to maintain the variety or inbred line. Breeder's seed is a class of seed in a seed certification program that is produced under the supervision of the plant breeder or owner of the variety. Basic seed (sometimes called foundation seed) is produced from breeder's seed and is the last step in the initial seed multiplications and is used to produce certified seed. Responsibility for producing basic seed often rests with seed production organizations, and the maize breeder must maintain supervisory involvement to ensure that the purity of the variety, inbred line, or hybrid does not deteriorate.

During the multiplication of breeder's seed and basic seed, strict quality control measures are essential. The goal is to selecting disease-free plants that have true-to-type seed with high germination. To achieve this, the fields must be effectively isolated to prevent open pollination with unwanted pollen sources, the controlled pollination phase must be carefully managed, and all off-type plants must be rogued.

Depending upon the quantity of breeder's seed needed and the amount of parental seed available, the quantity produced can be adjusted by increasing row number and row length in the seed maintenance plot. To maintain the highest possible level of purity, breeder's seed maintenance plots should be small, highly isolated, and rigorously rogued for off-types. It is best to multiply breeder and basic seed in areas where the variety is adapted and to keep the frequency of seed increases to a minimum. Shifts in the genetic make-up (and hence phenotypic characteristics) of open-pollinated varieties can occur rapidly when the seed is multiplied outside areas in which the variety is adapted.

By maintaining large reserve stocks of breeder and basic seed of popular genotypes, the problem of genetic drift can be reduced. Adequate reserve stocks are also vital as insurance against crop failure. Enough seed from the progenitors (parental material) of breeder's seed to last for at least three generations should be kept in safe cold storage. Similarly, reserves of breeder's seed and foundation seed should be adequate for at least two generations.

Certified Seed Production

Certified seed is the last stage in the seed multiplication process. It is generally produced from basic (foundation) seed. Certified seed should be grown in isolated plots and off-type and diseased plants should be rogued before pollination, although the percentage of roguing will be less severe than in breeder and basic seed plots.

Maize seed organizations employ various systems for certified seed production. Some seed enterprises prefer to engage directly in seed production, using their own farms and staff. Other seed enterprises select farmers to grow seed under contract with the guidance of their seed technologists. This system is advantageous. It requires less capital to begin operation, and, with proper production incentives, private seed growers will tend to manage their seed multiplication fields more carefully. However the seed enterprise must provide adequate technical support and supervision to ensure that the seed growers live up to their contracts. Having some limitation on the number of grower and some concentration in their location is also desirable to keep supervision costs reasonable.

In planning a certified seed multiplication and production programs, careful thought must be given to how each variety, inbred line, and hybrid will be maintained and stored, the number of multiplication cycles required to produce sufficient certified seed, and the amount of seed that must be retained to support later multiplications. Table 8.3 illustrates the requirements for multiplying breeder, basic, and certified seed in order to have enough certified seed of an open-pollinated variety to plant 200,000 hectares. If farmers use 20 kilograms of maize seed to plant 1 hectare, 4,000 tons of certified seed would be needed. The production of this quantity of certified seed requires 1,333 hectares to be planted with 20 tons of basic seed (at a seeding rate of 15 kg/ha). Twenty tons of basic seed, plus 20 tons of reserve basic seed, can be produced on an area of 20 hectares, using 300 kilograms of breeder's seed. That

TABLE 8.3 Seed certification program to produce sufficient open-pollinated variety seed to plant 200,000 hectares: Area, seed requirements, and production of different seed categories.

Seed category	Expected yield (t/ha)	Area (ha)	Seed rate (kg/ha)	Seed required (kg)
Breeder	1.0	0.6[a]	10	6
Basic (foundation)	2.0	20[b]	15	300
Certified	3.0	1,333	15	20,000
Commercial	–	200,000	20	4,000,000

[a] Sufficient for production of about 40 t (20 t of basic seed plus 20 t of reserve stock).

[b] Sufficient for production of 600 kg (300 kg of breeder's seed plus 300 kg of reserve breeder's seed).

amount of breeder's seed (plus 300 kg of reserve breeder's seed) can be produced from parental material grown on 0.6 hectare. Thus only three seed multiplication cycles are needed to go from a few kilograms of parental seed to the production of 4,000 tons of certified seed of an improved open-pollinated variety.

In comparison, producing seed of a double-cross hybrid takes five cycles of multiplication from breeder's seed to reach 4,000 tons of certified seed for commercial production (Table 8.4). The double-cross hybrid seed production system is clearly more complicated and costly.

Value Added in Seed Production

The additional costs (over the normal crop costs) of growing maize seed in developing countries and in the United States are shown in Table 8.5. These seed production costs vary according to the type of seed being produced. Single-cross hybrids are the most expensive to produce.

The production agronomy of a maize seed multiplication field and normal commercial grain field is quite different. In seed multiplication fields, planting densities are lower in order to achieve maximum ear development. Also seed production fields have to be isolated to prevent them from contamination by pollen from other maize genotypes. Other

TABLE 8.4 Seed certification program to produce sufficient double-cross hybrid seed to plant 200,000 hectares: Area, seed requirements, and production of different seed categories.

Seed category	Expected yield (t/ha)	Area (ha)	Seed rate (kg/ha)	Seed required (kg)
Breeder[a]				
Inbred A	1.5	0.35	10	3.5
Inbred B	1.5	0.15	10	1.5
Inbred C	1.5	0.15	10	1.5
Inbred D	1.5	0.05	10	0.5
Basic[b]				
AxB				
inbred A	1.5	11	10	110
inbred B	1.5	3	10	30
CxD				
inbred C	1.5	2.5	10	25
inbred D	1.5	0.75	10	8
Certified				
AxB	3.7	1,100	15	17,000
CxD	3.7	275	15	4,000
Commercial	–	200,000	20	4,000,000

[a] Seed multiplication area should produce a reserve stock equal to 2 to 3 year's consumption.

[b] Seed multiplication area should produce a reserve stock equal to 1 year's consumption.

additional costs include rouging, detasseling (hybrids), field supervision, and inspection. Because high quality maize seed costs considerably more to produce than grain, seed production activities should be undertaken in areas where soils and growing conditions are most favorable for the crop.

Hybrid seed production is even more expensive than open-pollinated variety production. Aside from bearing the cost of extra labor for detasseling and roguing, hybrid growers must learn more complicated field planting layouts and more exacting harvesting procedures. High level crop management skills are required to plant and grow separate rows of male and female plants in a way that will synchronize pollen shed from the male plants rows with silking and pollen reception in the female plant rows. Unless the cross-pollination phase is carefully managed and timed, the desired hybrid combination will not be obtained.

Seed Processing

The processing of certified seed is done to produce clean, disease-free seed with high germination and viability and uniform size (which is important for machine-planted seed). After harvest, the ears from each seed field (and grower) are grouped into separate lots and taken to processing plants. Here, the ears are dehusked and off-type or diseased ears are removed. Moisture and germination tests are conducted on each lot. If viability is good, the seed is dried until the moisture content is 12 percent or less. Seed that has higher moisture is more likely to deteriorate and spoil, especially in warm, humid climates. After drying, the ears are shelled and the seed is cleaned, sized, and tested for quality. Quality seed from comparable lots is then blended, chemically treated with pesticides, and packaged, normally in 25 and 50 kilograms bags. Processing, certification, conditioning, and packaging adds 30 to 50 per-

TABLE 8.5 Average variable seed production costs (US$/ha) over costs of high-yielding commercial grain production.

Item/activity	Developing countries		USA
	Improved variety	Double-cross hybrid	Single-cross hybrid
Parent seed	17	27	100
Fertilizer	100	100	100
Pesticides	37	42	70
Herbicides	20	23	60
Roguing	8	9	15
Detasseling	–	44	115
Supervision/inspection	20	24	15
Harvesting	36	49	75
Total	237	318	550

Source: CIMMYT1987.

cent to the value of commercial seed. The average cost per ton of the main processing activities in developing countries and the United States is compared in Table 8.6.

Seed Storage

When seed processing is completed, basic and foundation seed should be stored in a warehouse under cool, dry conditions. Storage facilities for breeder's and basic seed should be of higher quality than facilities for certified seed. Temperature control (through refrigeration) is especially important for these initial class of seed increase. Precautions must be taken to keep rodents and stored-grain insects from damaging seed in storage. Ordinarily, if storage conditions are good, certified seed can be stored for two seasons with little loss in germination viability. To hold down energy costs, intermediate-term storage of processed seed is best located in areas that have cool climates.

Seed Marketing and Distribution

Supplying farmers with the right kind and amount of seed at the right time is generally more difficult than producing quality seed, per se. Gross margins equal to 30 to 40 percent of the value added are needed to provide wholesale and retail seed distributors with sufficient profit to cover all seed marketing costs. A major cost is inventory finance. Because seed is a seasonal product, there is a considerable lag from the time when the seed is produced and processed and when the farmer buys it. In maize, this lag may be 6 to 12 months if the purchaser is a wholesale distributor. In addition the seed distributor often must extend credit further—to the farmer who may not be able to pay until after the harvest 4 to 6 months later. Over-production of a particular seed type can be very costly because it means that the seed distributor must either sell off his stocks at a considerable discount or carry the surplus seed inventory for another 12 months before being able to dispose of it. Field testing programs to demonstrate new varieties and hybrids to farmers add to marketing costs: Seed must be distributed to

TABLE 8.6 Selected costs in seed processing (US$/t).

Activity	Developing countries[a]	USA[b]
Transport	12	4
Drying seed	17	19
Certification	5	1
Conditioning/packaging	63	96
Total	97	120

[a] All types of maize seed. [b] Hybrid seed.
Source: CIMMYT 1987b.

farmers for testing along with technical advice, and advertising in farm periodicals, and training of seed dealers must be supported.

Seed Quality Control

If a maize seed production system is to succeed, it must consistently offer farmers better seed than they can produce themselves. Seed quality is judged in terms of trueness to type, germination percentage, cleanness (absence of undesirable inert material and seed of weeds or other crops), and freedom from seed-borne diseases. To achieve good quality, seed growers must take care to employ high-yielding crop management practices, to rogue off-type plants, and to harvest in a careful and timely way. During seed drying, correct temperatures and timing must be maintained to reach the desired moisture content (12 to 15%) without overheating that would lower germination. During processing, care must be taken to avoid admixtures, physical damage to the seed, and improperly applying chemicals against seed-borne diseases. During storage, seed lot identification must be maintained and the seed must be held under suitable climatic conditions to maintain high germination rates and prevent pest damage. During distribution, care is needed to keep the seed from being exposed to excess humidity or heat, to prevent contamination, and to maintain proper identity of the seed lot. Also, seed distributors must refrain from selling old seed stocks, which will invariably have lower germination. Such deceptions give commercial seed a reputation of poor quality, even though seed production and processing might have been done to the highest standards.

To help ensure maize seed quality, many governments use such measures as seed testing, seed certification, and general legislation on seed marketing. A seed testing laboratory is required to test for trueness to varietal characteristics and quality characteristics such as high germination percentage, cleanness of seed (freedom from debris, noxious seeds). A seed certification program depends primarily on the results from the seed testing laboratory to verify seed quality standards. It also is concerned with honesty in labeling information (type, weights, seed treatments, breeder, producer), determining eligibility of varieties, and various educational activities. To maintain objectivity, seed testing and certification programs should be independent of seed production and marketing programs.

The various components of a maize seed quality control system are typically introduced sequentially as seed industries become more mature. Care should be taken to avoid burdening a fledgling maize seed industry with excessive seed quality controls and bureaucratic regulations. In the final analysis, farmers are the ultimate judges of maize seed quality. They will not purchase seed from organizations that produce sub-standard quality.

Seed Industry Policy

Many types of organizations make up a maize seed industry. The activities and efforts of maize researchers, extension officers, consumer protection agents, seed producers and distributors, seed certification and testing agencies should be coordinated and synchronized. A consistent set of government policies and legislation are needed both to promote development and to regulate a dynamic maize seed industry.

National Seed Board

The establishment and regulation of a high quality seed industry is facilitated if there is a national seed board, with national subcommittees for the major crops (e.g., maize) and crop groups. The national seed board and its subcommittees should include representatives from public and private organizations that are engaged in agricultural research, seed production, seed certification, commercial farming, economic planning and policy making, and finance. The national seed board should wield considerable weight and authority (at least derived) in setting national seed industry policy on such issues as the roles of public and private organizations, importation and export of improved germplasm and commercial genotypes, and prices of public-sector germplasm.

Ownership of Seed Industries

Different forms of ownership exist in the maize seed supply systems in developing countries. They range from exclusively public-sector seed industries, to mixed public-private maize seed systems, to cooperatives, to solely private maize seed enterprises. Most developing countries, at least in launching national seed industries, have emphasized parastatal organizations to carry out seed production. Many of these public seed corporations (national and state or province) have had virtual monopoly control over seed of important food crops. This control resulted from explicit laws that control basic and foundation seed production and supply, thus reserving the right of seed production to the State, or from price controls and subsidies to the public seed corporation, which made it futile for private enterprises to invest in seed production.

Table 8.7 shows that government organizations and agencies are heavily involved in maize seed industries in developing countries. Of 48 countries, 40 reported having governments-run seed certification programs or seed testing laboratories, and 36 reported having government seed corporations that produce certified maize seed for commercial sale.

Sadly, the performance of publicly funded maize seed organizations in most developing countries has been disappointing. Despite substantial subsidies, few have been able to build and maintain strong demand for their seed products. The efficiency of public seed-producing organizations has been hampered by civil service rules and regulations, which result in overstaffing and poor work incentives. Their seed production and processing facilities are frequently too large for efficient operations. Consequently government maize seed organizations usually produce low quality seed at high cost.

The maize seed subsidies provided in public-sector seed organizations go mostly to the farmer directly (in the form of a very low price) and to the seed producer and processor (to compensate for inefficient operations). The seed distributor, however, is often heavily squeezed by this system of subsidies, receiving pressure from customers—the farmers—to respect official prices and from its inefficient supplier—the public-seed organization—which is trying to pass on its high cost of production. Most publicly funded maize seed systems provide inadequate profit margins (20 to 30%) to distributors and inadequate financing to maintain sufficient inventories. As a result, poorly supplied and sluggish seed distribution systems have discouraged farmer adoption of certified seed of improved varieties and hybrids.

In developing countries that have both public and private maize seed enterprises, private businesses produce two-thirds or more of all certified maize seed sold (Table 8.8). This is true even when a large public-sector maize seed organization exists (unless public-sector seed is so heavily subsidized that farmers cannot refuse it). Private enterprises usually focus on hybrid seed production, encouraged by the proprietary

TABLE 8.7 Public-sector involvement in maize seed industries.

Region	Countries reporting (no.)	Number of countries with		
		Seed regulation agency	Public seed organizations	Private seed companies
Africa	12	9	10	9
Asia	12	11	11	8
Latin America	13	10	11	12
All developing countries	37	30	32	29
Eastern Europe & USSR	4	4	4	0
Developed market economies	7	6	0	7
All countries	48	40	36	36

Source: CIMMYT 1987b.

nature of hybrid seed pedigrees and the farmer's need to buy the fresh seed each year. However, in some countries, notably Thailand, Egypt, and Guatemala, private seed enterprises still produce significant amounts of open-pollinated varieties (see Chapter 10).

Developing country governments are increasingly encouraging private participation in seed production as well as in the supply of other essential inputs such as fertilizers, pesticides, and machinery and equipment. Economics, not ideology, is motivating this change in national policies.

More private commercial seed enterprises are being established, both as national (usually with links to an international firm) and transnational business activities. During the 1990s, private maize seed production is likely to grow rapidly, as liberated economies release entrepreneurial talent capable of capitalizing on the large amount of improved germplasm available to serve as a catalyst in the modernization of maize production.

Seed Enterprise Scale and Investment Requirements

One of the virtues of a maize seed enterprise is that it need not be large to be economically competitive. Maize seed production is perhaps the only major agricultural input supply industry that does not require large investments in physical plant (such as for fertilizer or pesticide manufacturing). Even in the United States where 10 companies account for about 60 percent of hybrid maize seed sales, the remaining 40 percent is produced by several hundred seed companies located throughout the important maize-growing areas. Smaller hybrid maize seed companies generally rely on inbred lines developed by public-sector organizations or specialized private maize breeder enterprises that only develop inbred lines and new hybrid combinations. Most are profitable family-farm enterprises that produce less than 500 tons of hybrid seed annually, principally for sale to nearby farmers.

Economically viable maize seed industries also do not require com-

TABLE 8.8 Private-sector share of maize seed sales in non-centrally planned countries.

Region	Countries reporting (no.)	Private enterprises' share of total sales		
		Improved OPVs	Hybrid seed	All improved seed
Africa	11	57	95	92
Asia	11	62	62	62
Latin America	13	70	96	92
Developed market economies	7	100	100	100
All non-centrally planned countries	42	65	98	94

Source: CIMMYT 1987b.

plex and expensive seed processing equipment to produce a high quality product. Indeed, in developing countries a more common problem has been the installation of excessively large and sophisticated seed processing plants that are costly and difficult to operate and maintain. When a nation has only a few huge seed processing plants, seed stocks tend to get overly concentrated, and seed distribution becomes a problem, especially to maize-growing areas distant from the processing plant. It would be more efficient to have many smaller plants located in the appropriate geographic regions.

A maize seed enterprise that efficiently produces and sells 500 tons of certified maize seed (sufficient to plant 25,000 hectares of commercial maize land) each year can generate US$250,000 to $600,000 in gross income, depending on the seed type produced. Certified seed production can be accomplished by a cadre of contract growers who use their own land and equipment and are supervised by a small number of seed company agronomists. For 500 tons of seed output, small to moderate-scale, relatively inexpensive equipment can be used for seed drying, cleaning, classification, treatment, testing, and packaging to produce a quality seed product. Modest structures are needed to protect field and seed processing equipment, to store breeder, basic, and certified seed stocks, and to provide administrative office space. Finally, the seed enterprise needs operating capital to finance the inventory costs of seed stocks that are purchased and held from the end of one season to the following season before being sold.

Seed Pricing Policy

Many developing country governments exercise some control over maize seed pricing. Usually retail prices are set through negotiations between the seed industry and government. Some countries, however, control only the prices of seed produced by public enterprises, leaving private enterprises to operate as they see fit. Governments that take part in setting maize seed prices should apply sound commercial criteria. If official prices are set below the real costs of production, private capital and management expertise will not be attracted to seed production.

Governments in developing countries often subsidize maize seed production costs and retail prices even though there is little evidence that direct subsidies lead to an effective maize seed industry. A seed industry that depends upon subsidies—and not operating profits—for its economic survival is in a precarious financial situation, subject to the vagaries of the political budgeting process.

The price required to engage good seed growers will vary according to the types of assistance provided by the seed enterprise, the riskiness

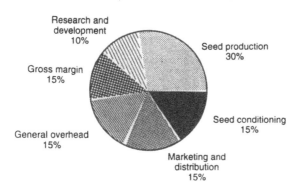

FIGURE 8.2 Typical breakdown of hybrid seed prices.

of seed production, and alternative investment opportunities. A premium of 15 to 20 percent above the price of maize grain is considered to be a minimum amount required to draw above-average maize farmers into the production of open-pollinated varieties (hybrid production requires greater premiums).

The average value-added in seed production, processing, and marketing varies considerably. Single-cross hybrids are the most expensive to produce. Improved open-pollinated varieties are the least expensive. The difference is due primarily to the lower seed yield in inbred-line multiplication compared with varieties and secondarily to the higher field operation costs in hybrid production. A typical breakdown of the price of hybrid seed in a private company shown in Figure 8.2 is useful for making cross-country comparisons of cost and price structures. The breakdowns of the sales prices for Brazil, Mexico, and India in Table 8.9 are remarkably similar although of course the absolute prices differ, reflecting the varying production costs in the three countries.

TABLE 8.9 Production and processing cost for double-cross hybrid maize seed sold by small private companies in Brazil, Mexico, and India, 1992.

	Brazil		Mexico		India	
	US$/kg	%	US$/kg	%	US$/kg	%
Basic seed[a]	0.18	16	0.20	12	0.08	14
Seed production	0.36	33	0.63	38	0.19	32
Seed conditioning	0.08	7	0.11	7	0.08	14
Total	0.62	56	0.94	57	0.35	60
Seed sales price	1.10	100	1.66	100	0.58	100

[a] Parent seed purchased by private seed companies from public-sector organizations. This cost can be considered part of the research and development investment for these companies.
Source: Adapted from CIMMYT 1994.

TABLE 8.10 Ratio of the price of maize seed to the price of grain in developing countries, 1992.

Region	OPVs	Nonconventional hybrids	Double-cross hybrids	Three-way-cross hybrids	Single-cross hybrids
Sub-Saharan Africa	4.9	6.6	6.8	5.2	6.1
West Asia and North Africa	3.8	—	8.4	8.6	16.1
Asia[a]	5.1	4.4	6.8	10.8	24.1
Latin America	5.4	—	10.3	26.3	23.3
All developing countries[a]	5.1	6.5	10.1	14.7	19.1

[a] Excluding China.
Source: CIMMYT 1994.

Maize seed is more costly than the grain. The seed grower has to be compensated for using proper production methods and for the additional risks incurred in certified seed production, and the costs of distribution have to be covered. Table 8.10 shows ratios of seed prices to commercial grain prices (in developing countries), based on extensive surveys routinely conducted by CIMMYT.

9

Improving Maize Technology
Delivery Systems

Maize is produced under a broad range of ecological and technological conditions around the world. At one end of the technology spectrum are the resource-poor, small-scale farmers of the low-income developing countries who use few external resources (inputs, hired labor, machinery). In these systems the productivity of the individual farmer is low and production of maize hardly exceeds the family's consumption. At the other end of the spectrum are the large-scale farmers, mainly in the industrialized countries, who use capital-intensive, science-based production technologies that are considerably dependent upon the use of external resources, especially machinery and agricultural inputs. In these systems, the productivity of the individual farmer is high and large maize surpluses are produced.

Until early in the 20th century, maize production technology was quite similar the world over. Yields were in the 1.0 to 1.4 t/ha range and production increases were achieved largely by expanding the area cultivated. Few external resources were used. The farmer relied on his own seed for planting and on barnyard manure, crop rotations, and fallows to maintain soil fertility. During the last 60 years, however, the rise of science-based agriculture has led to revolutionary changes in methods of maize production, especially in the industrialized countries, which have pushed grain yields continuously higher. During 1989-91, the average maize yield for the developed countries, excluding centrally planned economies, was 6.7 t/ha. The 1992 maize harvest in Iowa—the heart of the U.S. Corn Belt—was 45 million tons from 5 million hectares, for an average yield of 9 t/ha.

Unfortunately, the spread of modern maize production technologies in most low-income developing countries, especially those that have tropical and subtropical production environments, has been much less dramatic. In most regions, yields are still below 2 t/ha. Only North Africa (4.3 t/ha) and East Asia (4.6 t/ha) had average yields during 1990-92 that compared well with the world average (3.8 t/ha).

Average maize yields tend to conform to a climate/temperature gradient. Maize yields are highest in temperate environments, lower in the subtropical environments, and lowest in the tropical environments (Table 9.1). Although well-watered temperate zones have the highest maize yields, the yield differences between these production environments has more to do with the level of maize technology used than with the yield potential of the environment, per se.

Despite the superior economic yield potential of temperate zones, the theoretical yield potential of maize production is quite high in the tropics and subtropics (Plucknett 1992). Experimental yields of 8 to 12 t/ha are frequently obtained on research stations, and some maize growers in hot climates of Asia, Latin America, and Africa routinely obtain yields above 6 t/ha. Crop physiologists in the Netherlands who attempted to determine the "upper limit of what can be grown on all suitable agricultural land," taking into consideration sunlight, moisture, soil nutrients, number of days that crops can be grown, disease and insect pest matrices, etc., concluded that the highest theoretical grain production potential is found in Latin America, Africa, and Asia, not in North America, Europe, or Australia, where high-yielding modern agriculture prevails (Linneman et al. 1979). The key point is that significant maize-productivity gains are possible in the developing world, providing farmers have access to the necessary technological components, mainly improved genotypes and fertilizer.

Effectiveness of Maize Technology Delivery Systems

Maize technology delivery systems comprise public and private institutions and entities that develop and supply new genotypes, improved crop management practices, and production services to farmers. These entities include research organizations; extension and technical advisory services; suppliers of improved seed, fertilizers, crop protection chemicals, and farm machinery; and agricultural lenders.

The development of maize technology delivery systems is strongly influenced by government policies. In industrialized countries, the private sector plays a major role in the delivery of improved maize technologies to farmers. These private businesses have benefited greatly from publicly funded maize research and development programs. Over time, a well-integrated and highly effective public-private maize research and development system has evolved, one that is capable of generating and transferring to farmers a continuing stream of productivity-enhancing maize technologies.

TABLE 9.1 Estimated area and average maize yields for different environments in developing countries, 1989-91.

Environment	Area (million ha)	Yield (t/ha)	Estimated economic yield potential[a] (t/ha)
Tropical	34.7	1.4	4.0
Subtropical	19.1	1.7	4.5
Highland	6.2	1.6	5.0
Temperate			
Developing	22.2	4.2	6.5
Eastern Europe & USSR	10.9	3.5	6.0
Developed market economies	35.6	6.7	7.5
World	128.8	3.7	4.6

[a] Optimum yield potential at a point in time, given current input prices, availability of technology, and farmer management skills, as determined from on-farm trials.
Source: FAO Production Yearbooks and authors' estimates.

In most developing countries, the organizations that make up the national maize technology delivery systems are primarily within the public sector. Although public institutions have had success in developing new maize varieties and improved crop management practices, government organizations responsible for delivering improved technology to farmers generally have not functioned well. As a consequence, maize research pipelines are full of productivity-enhancing technological components that generally fail to get beyond the boundaries of experiment stations.

The efficiency of maize technology delivery systems can be gauged from the gap that exists between farmers' actual yields and optimum economic yields observed in on-farm maize technology validation trials (Table 9.2). Countries that have low efficiency ratings can achieve significant economic gains from improving maize technology delivery systems. Some gap between actual and potential yields, perhaps 15 to 25 percent, will always exist because of the lag between the generation of new research products and crop management information and the time it takes to deliver new technological components to the farm level. However, a national maize technology delivery system whose technical efficiency is below 50 percent has serious structural flaws. Low technical efficiency scores are generally caused by some combination of ill-functioning supply systems for seed and fertilizer, farmers' poor knowledge of recommended crop management practices, discriminatory price policies, and farmers' lack of sufficient capital to employ the recommended inputs and crop management practices (Byerlee 1987).

In considering the constraints to increased maize productivity in developing countries, a distinction must be made between countries in "pre-Green Revolution" phase and those in "post-Green Revolution" phases. In the pre-Green Revolution phase, farmers use few purchased

TABLE 9.2 Technical efficiency of technology delivery systems in major maize environments of selected countries.

Country	Environment	Economic yield potential[a] (t/ha)	Estimated actual yield (t/ha)	Technical efficiency (%)
United States	Temperate	9.0	7.5	85
China	Temperate	6.0	4.8	80
Pakistan	Subtropical	3.5	1.4	40
Thailand	Tropical	3.5	2.5	71
Philippines	Tropical	3.5	1.5	40
Egypt	Subtropical (irrigated)	6.5	5.9	90
Benin	Tropical	3.5	1.0	29
Ghana	Tropical	3.5	1.4	40
Nigeria	Tropical	3.5	1.4	40
Kenya	Tropical	3.5	1.4	40
	Highland	4.0	2.0	50
Zimbabwe	Tropical mid-altitude	4.0	1.8	45
Argentina	Temperate	6.0	3.5	58
Brazil	Tropical	3.0	1.4	47
	Subtropical	4.0	2.2	55
Ecuador	Tropical	3.5	2.0	57
	Highland	4.0	1.5	38
Guatemala	Tropical	4.0	2.0	50
	Highland	4.0	1.5	38
Mexico	Subtropical	4.0	2.5	63
	Highland	4.0	1.6	40

[a] Optimum yield potential, given current input prices, availability of technology, and farmer management skills, as determined from farmer-managed on-farm research trials.

Source: FAO and authors' estimates.

inputs such as improved seeds and fertilizers, and they rely on human and animal power and family labor for land preparation and crop cultivation. Raising yields under these circumstances involves diffusing improved varieties and fertilizer technology (mainly application of nitrogen) and promoting improved crop husbandry (especially weed control and better planting density). Adoption of improved technological components that can double or triple yields is feasible for most small-scale farmers, regardless of their education or resource endowments.

Moving the efficiency of the technology delivery system above 50 percent signifies entrance into a post-Green Revolution phase. In this phase, most farmers are using improved seeds and applying nitrogen fertilizers, although there is still scope in certain regions for further diffusion of these technological components. To improve technical efficiency in this phase that farmers must overcome complex second-generation crop management challenges. They must acquire better information and learn how to use a wide array of inputs efficiently. For example, improvements in soil fertility management (especially for secondary and micro-nutrients) and better timing and execution of critical cultural practices are essential to achieve further productivity gains. In addition, improved infrastructure is also required—especially road networks and marketing channels for inputs and output—to reach higher levels of technical efficiency.

Raising technical efficiency above 70 percent is probably only possible in a country that has a well-developed market economy and highly developed rural and agricultural infrastructure. Farmers in such countries produce maize strictly on a commercial basis. They must have access to current technical information, possess high level skills in crop management, and be able to operate close to the margin of economic efficiency. In countries with advanced maize technology delivery systems, all of the maize area is planted to high-yielding single-cross hybrids, which are grown under high levels of soil fertility and crop management. Advances in yields tend to be small and incremental. These highly efficient maize technology delivery systems generally involve mechanized field operations that, in turn, require a larger farm size and a well-developed agricultural credit system.

Farmer Knowledge of Improved Crop Management

In low-yielding indigenous maize-producing systems in the developing world, infertile soils place such a low limit on yield potential that little can be done to increase crop yields solely by manipulating cultural practices. Consequently, farmers prepare seedbeds haphazardly, resulting in patchy stands with poor spacing between plants. They neglect weed control because weeds are not highly competitive with the crop plant in infertile soils.

The move to science-based maize production systems requires traditional farmers to greatly expand their knowledge, technical skills, and managerial capacity. Improvements are needed in technical information transfer systems available to small-scale farmers. Farmers must manipulate and overcome a combination of biological factors constraining yields in an orchestrated and efficient manner. They have to employ precise land preparation, proper seeding rates and planting designs, and better conservation and management of soil moisture.

Farmers face increasingly complex fertilizer management requirements. After the adoption of high-yielding varieties, fertilizer use increases rapidly for several years. Usually, only nitrogen and phosphorus are applied. Later potassium and secondary and micro-nutrients may be needed to sustain high yields. Weed control is also very important because weeds become aggressive and highly competitive under improved soil fertility. Unless they are controlled mechanically or chemically, the farmer will harvest more weeds than food grains.

Similarly in low-yielding traditional maize production systems, insect pest and pathogen species, like the host plants, are all struggling for survival. But this situation changes dramatically in more-intensive maize production systems. Fertilized soils and improved agronomic practices result in the development of lush maize stands. Combined with year-round warm temperatures in most tropical and subtropical environments, the ecology within these fields becomes very favorable for development of disease pathogens and insects, and new methods of control.

Although crop management practices in traditional agriculture can be complex, much of the information is passed from farmer to farmer over generations. With the introduction of science-based agriculture, the value of the farmer's traditional knowledge is rapidly depreciated. Certainly the farmer has unequaled information about his land resources and the local climate. But his accumulated farming experience does not give him adequate knowledge of the appropriate timing and dosage of fertilizer application, nor of the need to plant high-yielding varieties densely, nor of the increased importance of early weeding once soil fertility is restored. This knowledge must be obtained from outside the farmer's traditional realm of experience.

Input Supply Systems

The history of agricultural modernization in the Third World has demonstrated that farmers, regardless of formal education or size of farm operation, can successfully manage improved seed-fertilizer production technologies. The use of these two essential inputs is estimated for 24 developing countries in Table 9.3. The most efficient maize technology delivery systems tend to be found in countries that have the highest use levels of improved seeds and fertilizers. For the most part, countries where few maize farmers use improved varieties and fertilizers are still in the pre-Green Revolution phase of agricultural development. In these cases, large yield gains can result from the diffusion and adoption of improved seed and fertilizer.

The challenge facing governments is to formulate policies that promote the development of effective input supply systems. In the past, governments have opted to develop publicly funded input supply organizations. With many failures, the trend now is toward shifting input supply functions to the private sector. The development of viable and dynamic private organizations will not, however, come overnight nor very easily. The high cost of rural transportation in many countries and the difficulty of reaching small-scale farmers with the needed technical information, as well as the inputs themselves, in a timely fashion argue

TABLE 9.3 Use of modern maize inputs in selected developing countries, 1989-91.

Country	Maize area (million ha)	Yield (t/ha)	Fertilizer use[a] (kg/ha)	Area (%) planted to Hybrids	OPV[b]
China	20.8	4.3	350	73	7
Brazil	12.1	1.9	50	63	7
Mexico	7.0	1.9	100	15	11
India	5.9	1.5	62	14	49
Philippines	3.7	1.3	20	5	9
Indonesia	3.0	2.1	43	3	27
Tanzania	1.8	1.4	40	11	7
Argentina	1.7	3.3	—	70	30
Thailand	1.6	2.5	9	15	84
Nigeria	1.6	1.3	30	5	25
Kenya	1.5	1.7	45	55	10
Malawi	1.3	1.1	33	5	2
Zaire	1.2	0.7	—	0	15
Zimbabwe	1.1	1.6	45	100	0
Ethiopia	1.1	1.7	35	6	10
Egypt	0.9	5.7	150	10	54
Pakistan	0.8.	1.3	80	3	23
Colombia	0.8	1.5	71	8	5
Zambia	0.8	1.9	75	45	1
Nepal	0.8	1.6	—	0	10
Côte d'Ivoire	0.7	0.7	—	0	10
Guatemala	0.6	2.0	50	36	24
Ghana	0.5	1.4	4	0	30
Turkey	0.5	4.0	180	33	13

[a] N, P_2O_5, K_2O.
[b] Improved open-pollinated varieties.
Source: CIMMYT 1992.

strongly for some continued public-sector developmental support in building effective agricultural input supply systems.

The importance of seed-fertilizer interactions should not be overlooked. Countries where farmers tend to plant improved maize germplasm but apply very little fertilizer can also realize productivity gains through greater use of fertilizers. Similarly, countries that use fairly high levels of fertilizer but little improved seed can raise yields through greater use of improved varieties.

Organization of Extension Services

The proper role and organization of agricultural extension services has been widely debated in recent years. In many countries the extension worker is a general government functionary engaged in broad activities including community development and nonfarm aspects of agricultural development such as public works supervision, assistance in electoral registration, public health campaigns, adult literacy campaigns, youth club management, etc. Although it is impractical to expect local extension officers to concentrate solely upon technical production matters, it is clear that they cannot be effective in supporting agriculture if they are at the call of several ministries or regional authorities.

It must be recognized that most extension field staff are poorly trained and that farmers often lack confidence in their technical and diagnostic abilities. In few developing countries does the field staff have a good secondary school education and at least 2 years of diploma training. Thus, upgrading the technical knowledge of extension officers inservice training courses and meetings is extremely important.

The training and visit extension system (T&V) promoted by the World Bank in the past decade seeks to strengthen three critical needs: the need to make extension staff into specialists in production technology, the need for sustained field efforts, and the need for regular instruction. The designers of the T&V system take the view that the organizational effectiveness of extension officers, most of whom are poorly trained, can be substantially enhanced by regular programs to upgrade technical skills, by concentration upon a narrow range of tasks, and by regular contact with farmers. To achieve these ends, an organizational hierarchy and management system are recommended with explicit job descriptions and chains of command.

A key feature of T&V is that extension personnel are limited to the diffusion of information. They do not distribute inputs or engage in other activities that could lead to conflicts of interest. Critics of T&V challenge this narrow focus, contending that most recommended technologies involve the use of improved varieties and fertilizers, which must be closely linked to input availability. The inadequacy of input supply organizations in many developing countries means that there is no practical alternative to involving extension staffs in some input supply work. For this reason, extension officers must help to administer input supply programs and assist with other agricultural services, such as helping farmers prepare applications for production credit.

In the T&V system, the individual farm visits are the primary method of conveying information to farmers and obtaining information on farmers' problems and requirements to feed back to the research service. A fixed schedule of visits to selected farmers is established for each extension agent to facilitate supervision and access to extension agents. Such visits normally take place throughout the growing season, often once every 2 weeks.

To support this program of farm visits, a regular system of meetings and training sessions led by supervising officers and specialists staffs must be held with groups of extension officers who operate in similar agricultural environments. Under most T&V programs, the meetings (usually twice a month) are devoted to instruction and explanations from subject-matter specialists and supervisory extension officers on the practices that should be recommended for various crops over the next 2 weeks.

Extension officers are unlikely to become effective technology transfer agents unless they are backstopped by specialists and unless there is government support for input supply services. One of the biggest single organizational constraints to establishing effective extension systems is scarcity of district-level subject-matter specialists who are sufficiently familiar with the relevant research work and possess adequate instructional skills for training local extension workers in the recommended technologies.

Also regular technical meetings are difficult to organize. Usually there are too few research staff and subject-matter specialists and they rarely are very mobile, instructional material is difficult to prepare and reproduce, and travel per diems have to be paid to extension agents who travel long distances. In short, simply arranging meetings on a regular basis across the country involves a major administrative effort and cost.

Although holding regular technical meetings is essential for effective extension work, it is not sufficient. Unless the extension agent is engaged with contact farmers in an active program of field demonstrations, training sessions and meetings with subject-matter specialists usually lack substance. And the production message that the extension agent takes back to the farmer also is not likely to be convincing. By incorporating an active and more realistic crop demonstration program, new life is breathed into the T&V technology transfer strategy.

Maize Production Campaigns

Successful maize production campaigns can have a positive training and motivational impact on extension officers. The ability to provide farmers with the means to radically improve maize productivity is a powerful morale booster for the extension service. With the production test plot as the common ground between the extension officer and the farmer, their relationship is much improved. When the extension officers work with farmers in the planting of the production test plot, bonds grow between them. The extension agents feel their work is more significant. The farmers gain respect for the extension worker. Farmers see their extension officer doing something tangible to improve the economic and social welfare of the community. The extension officer and the cooperating farmers begin to function as true change agents for rural development.

Maize production campaigns can accelerate the adoption of improved technology by farmers who use low-yielding production systems. These campaigns introduce modern inputs and production methods to farmers who have never bought fertilizers or commercial seed. They are most successful when they involve cooperation among research, extension, seed production, fertilizer supply, farm machinery, and agricultural credit organizations. The campaigns serve to focus economic and political attention on the productivity of the recommended technology, both for the individual farmer and the nation. Their goal is to produce dramatic changes in a short time. Maize production campaigns begin with basic improvements in production practices. If successful, subsequent extension efforts include advice on improved post-harvest storage methods, new varieties and hybrids, lower-cost and more optimum fertilizer use, improved weed and pest control, and new forms of mechanization.

Another major objective of maize production campaigns is to demonstrate research advances to policy makers. Through the campaign, policy makers are informed about the technology generation and transfer work of national research and extension organizations. They see the impact that the technology could have on farmers' fields, if adopted. They are made aware of the key components in the technology and the need to ensure that appropriate inputs are available to farmers on time and at affordable prices.

In the initial maize production campaigns, supply systems for these vital inputs may be weak. In the absence of functioning input supply systems, those leading the maize production campaign must take responsibility for producing the seed and supplying the fertilizer required for the field demonstrations. This informal method of seed production, bringing together the research/extension service and selected farmers, can become the forerunner of an organized seed production sector. Such informal seed production and distribution systems should not, however, be seen a permanent alternative to the development of a dynamic seed sector, which is needed wherever maize is a staple food.

Successful maize production campaigns involve many farmers and should be structured to influence political thinking. This dimension should not be neglected by the campaign planners. They are meant to be visible and their message is supposed to be dramatic. They are, after all, exercises in mass communication. Their purpose is to demonstrate major improvements in maize technology to as many farmers as possible. However, care must be taken not to over-emphasize maize production to the detriment of other crops that are important for the national economy.

Large Plots

Researchers and extension officers sometimes attempt to impress the farmer with new technology demonstrated in plots as small as 50 to 100 square meters, using statistics to extrapolate the small plot yield to the yield per hectare. The problem with such demonstrations is that the farmer often cannot visualize the extrapolation. The trial should be large enough to represent a legitimate test in the eyes of the cooperating farmer. The demonstration plot should be at least 0.2 hectare and preferably 0.4 to 0.5 hectare (a 0.4 ha plot would measure 65 m x 65 m) to have a strong psychological impact on the farmer. To contrast the yields of the traditional and recommended technology, the farmer should grow a companion plot along side—also a quarter to a half hectare—using traditional technology.

The Farmer's Role in Technology Diffusion

Farmer-managed field demonstration plots are the heart of an effective maize technology transfer campaign. These practical lessons are designed to let farmers test, evaluate, and possibly adapt the recommended crop technologies. Once the economic superiority of the recommended maize technologies has been verified in 30 to 40 test plots, it should be vigorously promoted through hundreds, and then thousands, of farmer-managed demonstration plots. As soon as the primary set of technological improvements has been accepted by substantial numbers of farmers, a new series of productivity-enhancing refinements should be introduced for testing, validation, and demonstration.

The participation of the farmer in these field demonstration plots is essential. S. A. Knapp, the founder of the U.S. Extension Service, stated the rationale for such plots almost a century ago, "What a farmer hears, he rarely believes; what he sees on someone else's field he often doubts; but what he does himself he cannot deny." This is still true today.

If a maize technology demonstration program persuades 10 percent of the farmers in a target area to adopt the recommended technology, then the technology is on the way to prevailing in the area. Usually, the demonstration program in each village should last 2 to 3 years. In the first year, a few demonstration plots (perhaps 10) should be established. Each plot will provide training for the participating farmer as well as for a "cluster" of neighboring farmers. In the second and third years, the number of demonstration plots can be expanded four- or five-fold.

Logistics of Technology Demonstration Programs

The extension service must be able to ensure that the farmers who take part in a maize demonstration and testing program have access to the improved seed, fertilizers, and other key inputs of the recommended production technology. For small-scale farmers in low-yielding maize environments, it may often be necessary to supply these inputs on loan, especially during the first year of field demonstrations in a village.

Requiring the farmer to pay for the inputs lent to him adds economic reality to the test plot. For practicality, it advisable for the participating farmers to repay their loans in cash. However, if improved open-pollinated varieties are being promoted and there is no reliable commercial source for the seed, some of the demonstration plots can also serve as artisan seed production plots, with the extension field staff accepting some of the best quality grain as in-kind payment for the inputs.

Extension Officer Preparation

A minimum number of extension officers are needed to cover a specified area and these front-line staff require adequate modes of transportation and modest amounts of field equipment to do their job effectively. The village extension worker should have a bicycle, or an off-road motorcycle, and a set of field equipment such as rubber boots, 50-meter tape, 100 meters of string, plot stakes, gunny sacks, a grain moisture meter, a field scale, and a plot data book with a water-resistant cover.

In most developing countries, village extension workers will need thorough training in the recommended crop management practices. The most effective instruction is "hands-on" training during the crop season. One-day training sessions will usually be sufficient. The initial training session should take place before planting, probably during land preparation, and should provide an overview of the entire crop cycle. The second training session should coincide with planting and the initial fertilizer application. The third session should occur at the initial weeding, 2 to 3 weeks after plant emergence, when the fertilizer side dressing is also applied. The fourth session should occur after flowering, when grain-filling is well advanced. The final session should occur at harvest, when yield data are obtained.

Farmer Training

Using the cluster system of participating farmers, the most intensive training in the recommended maize production technologies occurs during the first year. Farmer training sessions should mirror the training sessions for the extension agent: pre-planting, planting and initial fertilizer dose, weed control and second fertilizer dose, flowering and crop assessment, harvest, and yield calculation. The importance of providing proper farmer training, especially during the first year of demonstrations in a village, cannot be over-emphasized. It is expected that first-year participants will play a teaching and technology diffusion role with new farmers who join in the program in the second and third years. If bad weather, lack of inputs, or other unfavorable conditions occur at planting time, the extension officer should postpone the demonstration until a subsequent season. Field days should be organized for villagers to see the results of the demonstration plots and to discuss the factors responsible for the yields achieved.

Influencing Policy Makers

A dynamic on-farm maize demonstration program can have important impacts beyond teaching farmers about improved methods of maize crop management. In most developing countries where average maize yields are below 1.5 t/ha, delivery systems for improved maize seed and fertilizers are ineffective. Thus, the demonstration plot program must go beyond merely trying to convince farmers of the advantages of the improved technology. It must also serve to influence government policy makers and private entrepreneurs to take decisions that lead to the development of input supply systems and price policies that encourage farmers to try new practices. It is therefore vital that decision makers visit the demonstration plots to see first-hand the superiority of the production recommendations. The best time to invite influential people to visit the plots is during the field days organized toward the end of the crop cycle. Each extension officer should make sure that local government officials attend field days. Regional and national extension leaders should also induce higher-level policy makers and private entrepreneurs to attend field days as well.

Private-Sector Role in Technology Transfer

The role that private organizations can play in the development of the maize economy will change over time. In the early stages of development, private organizations must focus on the delivery of improved technologies and research products developed by public-sector organizations. Seed production and distribution of improved varieties and hybrids, fertilizer distribution, farm machinery services, and crop marketing are all prime candidates for private activity. As agricultural businesses expand and more capital is generated, greater opportunities emerge for research and development activities. Private seed companies can begin to develop proprietary hybrids and private input supply companies can begin to manufacture fertilizers and crop protection chemicals.

To get private investment capital to flow into the maize economy, several conditions must be exist. First, the physical and intellectual property of private enterprises has to be protected by the government from unlawful seizure or appropriation. Private entrepreneurs must be confident that their investments are secure and protected by impartial systems of law. Second, the government must ensure that no organization is granted special privilege that gives it an unfair advantage over other competing organizations. This means, for example, that public organizations engaged in the delivery of inputs must set prices for their products that reflect their true costs of doing business, rather than at subsidized levels. Third, private entrepreneurs who engage in agricultural business must have the potential to make a profitable return from their investments and managerial acumen and should not be denied or frustrated by government rules, regulations, and laws.

10

Maize Research and Development: Country Case Studies

This chapter contains case studies of the maize economy of six developing countries: Ghana, Zimbabwe, Thailand, China, Guatemala, and Brazil. These countries differ in their maize-growing environments, their research and development strategies for maize, and the importance and stage of development of their maize economies.

In each country, publicly funded institutions have been responsible for the bulk of maize research investments. Private maize research has been strongest in Brazil and Thailand, although it also is becoming more significant in Zimbabwe and Guatemala. CIMMYT, in particular, and IITA have made major germplasm contributions to national maize programs in Ghana, Thailand, Guatemala, and Brazil.

In all countries except China, private firms dominate the seed industry, although they often depend on varieties and hybrids developed by public institutions. Even in China, with its centrally planned economy, the maize seed industry is becoming more decentralized and adhering more closely to market principles. Thus, the trend in six countries is toward greater private-sector leadership in seed production and increasing private investments in proprietary maize research. Nevertheless, public maize research institutions still play an important role, especially in meeting the needs of small-scale, resource-poor farmers.

Ghana

At independence in 1957, Ghana was one of the richest countries in Africa. It possessed a fairly developed infrastructure, a relatively modern economy, significant foreign reserves, and no debt. It had the same per capita income as South Korea. However, in the first two decades after independence, the economy deteriorated. In 1983 Ghana launched a program intended to improve macroeconomic management and bring about structural adjustment in the economy. While the economy did not recover instantaneously, the World Bank and various development authorities believe that it is on the mend.

Ghana's Agriculture

Agriculture is central to Ghana's economic life. Seven and a half million people—50 percent of the population—are engaged in agriculture, tilling 3 million hectares of land. Agriculture provides about 50 percent of the GDP and more than 90 percent of this production occurs in the traditional sector where the average holding is less than 1.6 hectares. Arable and permanent cropland account for only 12 percent of the total land area, indicating that there is considerable scope for expanding the area devoted to food production. Little agricultural land is irrigated.

Ghana's agricultural economy is not highly diversified. Cocoa, the main export crop, covers about 900,000 hectares. Roots and tubers—cassava, yams, and cocoyams—are the major food crops and are grown on 700,000 hectares. Maize, sorghum, and millet—the major cereal grains—are collectively grown on 1 million hectares. Maize accounts for 547,000 hectares, and sorghum and millet occupy 254,000 and 192,000 hectares, respectively, in the drier northern regions of the country.

The National Maize Economy

Maize was introduced by the Portuguese traders to the coastal areas of Ghana (Gold Coast) during the 16th century. By the 18th century it had penetrated the forest zones, and by the 19th century it had spread

Ghana's agriculture in brief (1991).	
Population	15.3 million
Population growth, 1991-2000	3.2 %/yr
GDP	US$6,413 million
Per capita GDP	US$400
Agriculture vs GDP	53%
Adult literacy	60% total (women: 51%)
Labor force (1990)	51% in agriculture
Total land area	23 million ha
Arable & permanent cropland	2.9 million ha
Cereal crop area	1.0 million ha
Irrigation	22,000 ha

Sources: World Bank 1993; FAO 1993.

into the Guinea savanna zones. Over time, maize has replaced sorghum and millet as the paramount cereal. Today it covers 55 percent of the total cereal area. During 1990-92, Ghana produced 688,000 tons of maize and the average yield was 1.3 t/ha. About 95 percent of the maize grown in Ghana has white grain. Eighty-four percent is consumed as a human food; only about 11 percent is used as animal feed. Most maize is intercropped, primarily with cassava or cocoyam in southern Ghana or with sorghum, millet, or groundnuts in the Guinea savanna ecologies of the northern area. Small-scale farmers account for 90 percent of national maize production. Maize is consumed as fermented and steamed maize dough, porridges, gruels, and boiled or roasted green ears. Roughly half the total production enters commercial markets. The rest is consumed on the farm. Most farmers sell their maize to traders who come to the farm gate or to a trader at a nearby market.

During the 1950s and 1960s, maize production in Ghana increased as the area under cultivation expanded. Yields were stagnant. During the 1970s, reflecting the nation's economic crisis, maize production fell significantly. Both the total maize area and average yields declined. During the 1980s, the inception of dynamic national maize research and technology demonstration programs brought about a dramatic change in the maize economy. Farmers have planted a third of the total maize area to high-yielding maize varieties that have streak resistance. Extension workers also have helped farmers to improve their crop management practices, especially soil fertility, planting patterns, and timeliness of weed control. The result is evident in national maize statistics. Between 1983 and 1992, maize production increased by 8.3 percent per year, driven by a 8.1 percent annual growth in yields.

Ghana's maize economy in brief (1990-92).

Maize area		0.52 million ha
Maize yield		1.4 t/ha
Maize production		0.69 million t
Per capita utilization		45 kg/year
Net maize imports		15,000 t/yr
Maize grain color		95% white
Maize used as	human food	79%
	livestock feed	6%
Area planted to	unimproved varieties	65%
	improved OPVs	35%
	hybrids	0%

Source: CIMMYT 1994.

Principal Maize Environments

Maize in Ghana is grown in a tropical environment within which five production zones can be identified (Fig. 10.1). Maize is produced during both major and minor seasons in areas with bimodal rainfall patterns and during the major season in areas with unimodal rainfall patterns.

Occasional to frequent moisture stress characterizes all production zones (Table 10.1). Among the biotic stresses, maize streak virus and southern rust are the most pervasive disease problems. Termites and several species of maize borers are serious pests. In recent years, striga, a parasitic weed, has become increasingly damaging in the coastal savanna and Guinea savanna ecologies.

Coastal Savanna. The coastal savanna ecology is characterized by low and erratic rainfall (less than 600 mm), considerable cloud cover (low solar radiation), and sandy soils with low water-holding capacity. The rainfall pattern is unimodal, and planting occurs in April-May. Because of the demand for maize in Accra, Tema, and other coastal cities, 75,000 hectares of maize is planted during the major season. Yields are generally low and highly variable, depending upon the season's rainfall.

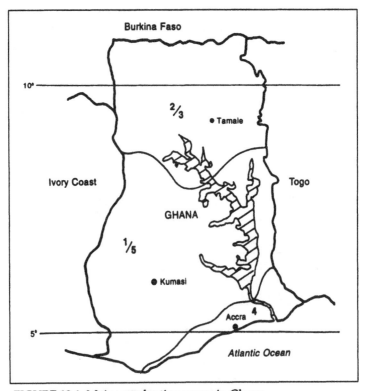

FIGURE 10.1 Maize production zones in Ghana.

TABLE 10.1 Characteristics of maize production zones in Ghana.

	Production zone				
	1	2	3	4	5
Area, 000 ha	200	100	25	100	100
Environment	tropical	tropical	tropical	tropical	tropical
Grain type	white dent	white dent	yellow flint	white dent	white dent
Growing season	major	major	major	major	minor
Maturity	late	intermediate	early	early	early
Moisture stress	sometimes	frequent	sometimes	frequent	frequent
Soil type	normal	normal	normal	normal	normal
Biotic stress[a]					
Streak virus	●●●	●●●	●●●	●●●	●●●●
Southern leaf blight	●	●	●	●	●
Fusarium stalk rot	●●	●	●	●●	●●
Charcoal rot	●	●	●	●	●
Fusarium ear rot	●●	●	●	●●	●●
Diplodia ear rot	●	●	●	●	●
Diplodia stalk rot	●	●	●	●	●
Southern rust	●●●	●●●	●●●	●●●●	●●
African maize stalk borer	●	●	●	●	●●●
Sesamia botanephaga	●●	●	●	●	●●
Eldana saccharina	●	●	●	●	●●
Termites	●●	●●	●●	●	●
Striga	○	●●	●●	●●	○

[a]Stress ratings: ○ Biotic stress not present in region. ● Present but not of economic importance. ●● Some economic losses. ●●● Significant economic losses. ●●●● Severe economic losses. ●●●●● Maize cannot be grown unless a resistant variety is grown or chemical control is applied.
Source: CIMMYT.

Humid Forest. In the humid forest ecology, average annual rainfall is above 1,400 millimeters and distribution generally is bimodal, leading to two planting seasons. Approximately 50,000 hectares of maize is grown in during the major season and 75,000 hectares during the minor season. The major season begins in March-April with harvesting in June-July; the minor season planting begins in August-September with harvesting in November-December. During 7 humid months, precipitation exceeds potential evapotranspiration. Soils are slightly acidic. Maize is frequently intercropped, particularly with cassava or yams.

Forest-Savanna Transition Ecology. The forest-savanna transition ecology, which spans forest and moist savanna ecologies, is the most important area for maize production. Because most of the forest has been cleared, much of the transition zone is savanna-like. Rainfall averages 1,000 to 1,400 millimeters per year and is bimodal to unimodal changing to unimodal toward the Guinea savanna. Roughly 200,000 hectares are planted to maize in the major season and 50,000 hectares in the minor season.

Guinea Savanna. In the Guinea savanna, approximately 75,000 hectares are planted to maize. The rainfall pattern is unimodal, and the single planting season begins in April-May and ends in July-August. The zone experiences 4 to 5 humid months per year. Solar radiation during the brief maize-growing season is high. The long dry season limits the incidence of pests and diseases. The Guinea savanna can be subdivided, according to moisture regime, into the "moist" and "dry" savanna. Maize varieties with very early maturity and drought-stress tolerance are desired by farmers.

The Maize Research System

Maize research is conducted by the Crops Research Institute, which is part of the Council for Industrial and Scientific Research under the Ministry of Science, Industry and Technology. In 1979, the Ghana Grains Development Project was established to improve maize and grain legume production. The Crops Research Institute and CIMMYT are the major executing organizations for this project, which is funded by government of Ghana and the Canadian International Development Agency. Under this project, CIMMYT has posted one to two maize scientists and IITA has posted one cowpea scientist. Other agencies taking part are the Ministry of Agriculture and the Grains and Legume Development Board (GLDB).

The Ghana Grains Development Project has promoted a broad strategy for maize research and development. Maize germplasm development and basic agronomic research are carried out by a cadre of well-

TABLE 10.2 Maize varieties released by the Crops Research Institute, Ghana.

Variety	Grain color	Maturity	Germplasm origin
Golden Crystal	yellow	110	Local
Composite-2	white	120	Local & Mexican/Caribbean
Mexican-17	white	120	Local & Mexican/Caribbean
La Posta	white	120	Pop 43, CIMMYT
Composite-4	white	120	Local & Mexican/Caribbean
Composite-W	white	120	Local & Mexican/Caribbean
Kawanzie	white	95	Tocumen (1) 7931, CIMMYT
Dobidi	white	120	Ejura (1) 7843, CIMMYT
Safita-2	white	95	Pool 16, CIMMYT/IITA
Dorke	white	100	Pool 16-SR, CIMMYT/IITA
Aburotia	white	105	Tuxpeño PBC16 (Pop. 49), CIMMYT
Okomasa	white	120	EV8343-SR, CIMMYT/IITA
Abeleehi	white	105	Ikenne 8149-SR, CIMMYT/IITA
Obatanpa (QPM)	white	105	Pop 63-SR, CIMMYT/IITA

Sources: Crops Research Institute and CIMMYT.

trained Crops Research Institute scientists. Station-based research is closely connected to an extensive program of on-farm research by the staffs of the Crops Research Institute and GLDB. The on-farm research is also linked to a program of technology demonstrations carried out by GLDB and extension officers.

The Crops Research Institute (CRI) has developed a number of improved open-pollinated maize varieties in collaboration with CIMMYT and IITA (Table 10.2). These improved varieties correspond to three maturity periods: late (120 days), intermediate (110 days), and early (100 days). Research on hybrid development is also under way.

Three recently released varieties (Okomasa, Abeleehi, Dorke) are versions of earlier improved varieties (Dobidi, Aburotia, Safita-2, respectively). These varieties and Obatanpa, another new release, have resistance to maize streak virus incorporated from IITA germplasm and have been improved for other characteristics, especially husk cover. In 1991 CRI released its first quality protein maize (QPM) variety, Obatanpa (which means mother's milk in Ashanti), based on CIMMYT QPM population 63. This intermediate-maturing variety (110 days) is resistant to maize streak virus and is encountering considerable approval among farmers.

CRI's improved varieties have been favorably received. Maize farmers have shown a preference for the intermediate-maturity varieties, such as Abeelehi or Obatanpa. These 105-day materials are high yielding, yet they mature 15 days earlier than full-season varieties such as Okomasa. The shorter maturity gives the farmer more flexibility in planting dates, which is especially important in areas that have erratic rainfall.

The maize production recommendations developed by CRI have dealt with selection of the appropriate variety, fertilizer use, plant stand management, and weed control. Considerable research to improve the predominant maize-based intercropping patterns has also been carried out. Good adoption of many technological components has occurred.

Research-Extension Linkages

Ghana's public agricultural institutions strive to link on-farm research with extension activities. Research officers responsible for managing on-farm experiments are also in charge of planting demonstrations in farmers' fields and using these plots as the basis of field days and as sites for in-service training of extension workers. Since 1986, the Sasakawa-Global 2000 technology transfer project has been active in Ghana, working with the Ministry of Agriculture to promote CRI's recommended maize technology through a extensive program of farmer-managed, half-hectare demonstrations called production test plots.

Sasakawa-Global 2000 provided some financial assistance for input distribution to farmers growing production test plots, travel and transport per diems, field equipment, and in-service training in the recommended production technologies for extension staff and farmers.

In 1991 with funds from a World Bank loan, the Ministry of Agriculture took steps to strengthen the extension service. Today, the Ghanaian extension service operates the unified extension development program based the training-and-visit approach and the large-plot demonstration programs of Sasakawa-Global 2000. Extension staff receive in-service training in improved maize production technology and maize-based cropping systems from subject-matter specialists trained by the Crops Research Institute, the extension service, other Ministry of Agriculture departments, especially crop services, and faculty of the agricultural universities. These front-line staff, in turn, make periodic visits to individual contact farmers as well as managing 5 to 10 maize production test plots in cooperation with farmers in the neighborhood. Four to five field days are held during the maize growing season to offer training to the farmers who have production test plots as well as to their neighbors.

Maize Seed Industry

In 1979 the Ghana Seed Company (GSC) was established as a public enterprise. For more than a decade it produced and distributed seed of improved crop varieties developed by the Crops Research Institute. Although GSC received considerable investment support from international aid agencies to build processing plants and seed storage facilities and to send staff for advanced training in seed technology, the company failed to operate at a profit. It produced small quantities of maize seed (it never sold more than 500 tons per year), which was frequently of poor quality (low germinating). Because of GSC's poor performance, GLDB produced 50 to 250 tons of commercial seed of improved maize varieties annually from 1983 to 1991.

In 1989, the government of Ghana enacted a new national seed development policy, which called for the private sector to take over commercial seed production and marketing of public-sector varieties. GSC was closed and its facilities were put up for sale. Under the new scheme, the Crops Research Institute is responsible for maintaining sufficient "breeders' seed" stocks of its registered and recommended varieties. GLDB is responsible for "foundation seed" multiplication. Private growers are responsible for commercial seed production for sale to farmers. And the Ministry of Agriculture's seed inspection unit is responsible for seed certification. A National Seed Service has been formed to coordinate the various public-sector activities in foundation seed production and commercial seed inspection and certification carried out by the Ministry of Agriculture and GLDB.

In 1991, the Seed Growers' Association (SGA) was formed to serve as a voice for private seed growers and to help build their capacity as viable private enterprises. In 1990-91, private seed growers produced about 400 tons of maize seed that was acceptable for certification. However, because of bottlenecks in seed processing, bagging, and marketing, only about 250 tons entered market channels. Within 3 years, SGA producers had nearly 800 tons of certified seed on the market.

Maize Production Constraints

The most important production constraint in Ghana is infertile soil. Traditional bush-fallow systems for restoring soil fertility have broken down. Previously the average fallow lasted 10 years, but today nearly three-fourths of the fields are fallowed for less than 5 years, and many fields are continuously cropped.

With the removal of government subsidies on fertilizer between 1988 and 1990, the real (deflated) price of fertilizer doubled. In 1990-91, the ratio between the farm-level price of nitrogen and the price of maize grain was about 6:1. Thus, it is not likely that farmers in Ghana will substantially increase fertilizer use on maize unless higher prices of maize grain restore a nitrogen-maize price ratio of 4:1 or lower. Until then fertilizer use on maize will be concentrated in the more favored production areas, which have more assured rainfall and relatively better soils.

Research is urgently needed to develop soil fertility management systems based on the use of inorganic and organic fertilizers. One promising green manure crop is velvet bean (*Mucuna utilis*). This grain legume can add significant amounts of nitrogen to the soil as well as help to control the noxious weed speargrass (*Imperata cylindrica*). Farmers will also need more cost-effective methods of weed control if they adopt higher yielding crop management practices that improve soil fertility. In the highest productivity zones for maize cultivation—the forest transition and savanna areas—declining soil fertility is resulting in increasing invasions by the parasitic weed striga.

Ghana's shortage of facilities for storing and marketing maize grain results in considerable price fluctuations as well as grain quality problems. Stored maize, other than the small amounts of shelled grain kept in containers in homes, often suffers significant damage from insects and various types of molds along with aflatoxin problems.

Sasakawa-Global 2000, working with the postharvest development unit of the Ministry of Agriculture, is mounting a major extension campaign to encourage small-scale farmers to improve their on-farm storage of maize grain. In addition, a more extensive regional network of commercial maize grain storage facilities—private and public—is being developed to serve maize farmers and consumers.

Prospects for the Maize Economy

Superior production technology has been developed in Ghana that is suitable for use by small-scale farmers who produce most of the maize. Farmers have shown that they are willing to adopt modern technological components, provided that the requisite technological components are accessible nearby and sufficient economic incentives exist.

The government of Ghana now looks to the private sector to develop input-supply industries. Over the long run, this strategy should lead to more efficient input-delivery systems. In the short run, however, it is not likely that the private sector will invest sufficiently in input-supply systems to serve small-scale farmers without some additional incentives from the public sector. Such incentives can take several forms. One would be for public institutions to provide loans or loan guarantees to encourage private entrepreneurs to enter agri-businesses. Another form would be to provide some level of subsidy to make fertilizer use more economically attractive for small-scale farmers.

At present, price instability discourages greater use of purchased inputs. During the harvest period, grain prices can drop to one-fourth of their level 2 to 3 months earlier. Yet farmers are forced to sell a portion of their maize immediately after harvest to meet various economic obligations (family needs, repayment of input loans). The grain farmers retain on the farm to sell later when prices are higher often suffers significant postharvest damage. Thus, improved on-farm grain storage systems are needed to allow farmers to safely hold grain for 5 to 6 months.

During the 1990s, growth in the consumer demand for maize in Ghana is projected to be slightly less than the population growth rate of 3 percent per year. However, the potential exists for increased demand for maize as an animal feedstuff and as a raw material in food-processing industries for beer-making and dry milling. If the government were to ban the import of malting barley and sugarcane—the principal carbohydrate sources currently used in domestic beer-making—and substitute domestic maize as the raw material (as Nigeria has done), the beer industry's demand for maize could reach 100,000 tons within the decade. In addition, demand for maize as a feed grain could also increase if an intensive poultry and swine industry begins to develop. Overall maize demand might grow by 5 to 7 percent per year during the 1990s, suggesting that to maintain self-sufficiency, Ghana's maize production would need to reach 1.2 to 1.5 million tons by the 2000, compared with 0.7 million tons in 1989-91.

To meet this projected demand, Ghana must seek to increase yields and expand the total maize area. Raising yield will require effective input-supply industries and adequate price incentives and price stability. To enable small-scale farmers to cultivate larger maize areas, they must have better access to animal and machine power.

An efficient agricultural extension system can contribute greatly to the modernization of the maize economy. Although the extension service has been strengthened in recent years through various grants and loans from the World Bank, IFAD, and the Sasakawa-Global 2000 program, it still has a considerable way to go to become a truly effective technology-transfer organization. In particular, training of extension officers must be strengthened and front-line staff should be provided adequate transportation and sufficient operating funds to mount dynamic field demonstration programs.

Zimbabwe

In the late 1950s and early 1960s, British decolonization was progressing relatively smoothly in most of Anglophone Africa. In 1965, however, 250,000 white settlers in Rhodesia, declared themselves the independent government of the country and its 6 million African inhabitants. After 12 years of defiant resistance and a war of liberation, the white government of Rhodesia conceded the inevitability of black majority rule. The changeover, which took place in 1980, was marked by the adoption of a new name for the country, Zimbabwe.

Zimbabwe's agriculture in brief (1991).

Population	10.1 million
Population growth, 1991-2000	2.3%/yr
GDP	US$5,543 million
Per capita GDP	US$650
Agriculture vs GDP	20%
Adult literacy	67% total (women: 60%)
Labor force (1990)	24% in agriculture
Total land area	39 million ha
Arable & permanent cropland	2.5 million ha
Cereal crop area	1.9 million ha
Irrigation	216,000 ha

Source: World Bank 1993; FAO 1993.

Zimbabwe's Agriculture

At the time of independence, Zimbabwe had a relatively well-developed agricultural system, with important export outlets for tobacco, cotton, coffee, and maize. The civil war during the 1960s and 1970s was an impetus for diversification because economic sanctions levied by most nations forced the country to produce a wide range of agricultural products.

Agriculture employs more than two-thirds of Zimbabwe's 10 million people and accounts for 32 percent of national exports. The mining industry (gold, asbestos, nickel, iron ore, ferro-chrome, coal, and copper) accounts for 35 percent of national exports. Maize is the major food crop, covering two-thirds of the cereal crop area. Millet and sorghum are important food crops in the drier regions of the country. Cotton and tobacco, the major cash crops, are important export earners.

A significant feature of the agricultural sector in Zimbabwe is its division into commercial and communal subsectors. The commercial subsector includes 5,400 large-scale farmers occupying 16 million hectares (some of these farms have been acquired for resettlement or are inactive) owned under the freehold system. They are found in the so-called large-scale commercial farming area. In addition, there are small-scale commercial farming areas, occupied by 9,000 black farm families on 1.5 million hectares. Most of this area is also under the freehold system.

The government of Zimbabwe has sought to redress the imbalance between land allocated to whites and blacks. To date, 600,000 hectares of former commercial farm land and 150,000 hectares of state land have been used to resettle communal area families. In 1990, the communal subsector was made up of 174 separate communal areas totaling 16.4 million hectares. These areas, collectively known as the communal lands

Zimbabwe's maize economy in brief (1990-92).	
Maize area	1.04 million ha
Maize yield	1.3 t/ha
Maize production	1.31 million t
Per capita utilization	133 kg/yr
Net maize imports	nil
Maize grain color	93% white
Maize used as human food	67%
livestock feed	19%
Area planted to unimproved varieties	0%
improved OPVs	0%
hybrids	100%

Source: CIMMYT 1994.

(formerly tribal trust lands), are home to 900,000 farm families. Land is held under traditional land tenure rights. Most of the communal lands are found in marginal zones where rainfall is erratic and often insufficient, and summer temperatures are high.

The National Maize Economy

Although maize was brought to the coast of Africa in the 16th century by Portuguese traders, it probably did not penetrate the region of Zimbabwe until late in the 19th century, with the spread of British colonialism. The colonial government began encouraging maize production for export at the beginning of the 20th century. By World War I (1914), nearly two-thirds of Zimbabwe's maize crop was exported, mostly to Europe for use as a livestock feed. During the 1930s, Zimbabwe's maize exports dropped significantly, as a result of the great economic depression, but rose again during the 1940s in response to the demand generated by World War II. By the 1960s, a large-scale commercial farming sector had developed that was using hybrids, fertilizers, and machinery. The Grain Marketing Board, an efficient parastatal, disposed of maize production, selling to private millers for maize meal and handling exports to other countries.

Today, maize accounts for roughly 70 percent of the total cereal crop area in Zimbabwe. Normally national production is 1.8 to 2.0 million tons and yields average between 1.5 and 2.0 t/ha. During 1990-92, however, 1.3 million tons of maize were produced annually on approximately 1.0 million hectares, with national yields averaging about 1.3 t/ha, as a result of Zimbabwe's worst drought in this century.

In Zimbabwe, in comparison with other countries of sub-Saharan Africa, small-scale farmers adopted improved maize production technol-

ogy during the 1980s at a remarkable rate. Maize production in communal areas has more than doubled as the government has brought
modern services to the hitherto-neglected small-scale farming sector.
With small-scale farmers rapidly adopting hybrids and increased fertilizer use, average maize yields have increased by 70 percent. While the
greatest yield and production increases have occurred in the higher
rainfall areas, technology adoption has been strong even in the lower
rainfall areas.

Principal Maize Ecologies

Zimbabwe is dominated by a large plateau in the central and northern parts of the country where elevations are generally above 1,200 meters. The country is often divided into three broad natural environments: the highveld, the area on the plateau; the lowveld, the areas of
lower elevation in the north, south, and southeast of the country; and
the midveld, the transitional area between these major environments.
On the basis mainly of elevation and annual rainfall, the country is divided into five production zones (Fig. 10.2). Virtually all of Zimbabwe's
maize-growing area lies in mid-elevation areas with a subtropical environment. There are, however, important differences in moisture availability (Table 10.3).

The commercial farming areas are found mainly in production zones
1 and 2, and the communal farming areas are found mainly in production zones 3 and 4. However, since independence, the government land

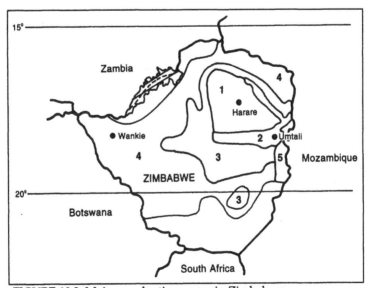

FIGURE 10.2 Maize production zones in Zimbabwe.

TABLE 10.3 Characteristics of maize production zones in Zimbabwe.

	Production zone				
	1	2	3	4	5
Area, 000 ha	350	400	350	260	40
Ecology	subtropical	subtropical	subtropical	subtropical	tropical
Grain type	white dent	white dent	white dent	white dent	white dent
Growing season	major	major	major	major	major
Maturity	late, extra late	late	int.	int.	int.
Moisture stress	sometimes	sometimes to often	often to usual	usual	often to usual
Soil type	normal	normal	normal	normal	normal
Biotic stresses [a]					
Northern leaf blight	●	●	●	●	●
Common rust	●	●	●	●	●
Ear rot	●●	●●	●●	●●	●●
Streak virus	●●	●●	●●	●	●
African maize stalk borer	●●	●●	●●	●	●
Spotted stem borer	●	●	●	●●	●●

[a]Stress ratings: O Biotic stress not present in region. ● Present but not of economic importance. ●●Some economic losses. ●●● Significant economic losses. ●●●●Severe economic losses. ●●●●● Maize cannot be grown unless a resistant variety is grown or chemical control is applied.

Source: CIMMYT.

reform program has supplied land in production zones 1 and 2 to communal farmers who now plant 200,000 hectares of maize there.

Roughly half of the maize-growing area of Zimbabwe suffers frequent to chronic drought stress. Intermediate-maturing maize is required for these areas. Biological stresses include ear rots, maize streak virus, and two stem borers, *Busseola fusca* and *Chilo partellus*.

The Maize Research System

Maize research has a long and successful history in Zimbabwe. The maize breeding program, which started in 1932, is second only to that of the United States in longevity. The original breeding program was based on developing open-pollinated varieties. In 1950, Southern Cross was the most common improved open-pollinated variety grown in the country. Since then, a succession of double, three-way, and single-cross hybrids has been developed from this germplasm pool. The single-cross hybrid SR 52, released in 1960, was one of the world's first single-cross hybrids to go into commercial production. SR 52 is a late-maturing (157 days), white dent genotype that has remarkable yield stability and adaptability to a wide range of agroecological conditions. It remained the most widely grown hybrid in Zimbabwe and much of southern Africa for more than 25 years, and it still covers 15 percent of the total maize area, more than 30 years after its release.

The foundation of Zimbabwe's present research structure was laid in 1948 with the establishment of the Department of Research and Specialist Services (DR&SS) following a major re-organization within the Ministry of Agriculture and Lands. During the colonial period, DR&SS, served the large-scale white commercial farmers and neglected the technological needs of the small-scale black farmers. Consequently, the gap between the productivity levels of the two farming communities widened steadily. However, one notable result of spillover colonial research was the introduction of hybrid maize in communal areas during the 1970s. By 1980, maize had become the major crop grown by small-scale farmers in Zimbabwe.

The agricultural policy in Zimbabwe after independence sought to redress the inequitable access to resources that had characterized the nation's past. Of particular importance the government's set out to increase agricultural production from small-scale communal farmers, while maintaining and even increasing production in the large-scale commercial sector. This had profound effects upon DR&SS. From catering to the research needs of fewer than 8,000 large-scale, commercial farmers before independence, DR&SS is now required to generate viable agricultural technologies for nearly 1 million resource-poor farmers in the communal areas while at the same time sustaining the productivity of the large-scale commercial producers.

Today, maize improvement research in Zimbabwe is carried out by a mix of public and private national and international maize research organizations. Collectively these institutions provide Zimbabwe with one of the strongest maize research programs in Africa.

Public-sector Maize Research. The Crop Breeding Institute (CBI) conducts maize breeding research at its headquarters at the Agricultural Research Center in Harare. In addition CBI conducts maize research at the Gwebi variety testing center, the Seed Co-op's Rattray-Arnold research farm, the lowveld research stations, and numerous other small testing sites scattered throughout the maize-growing areas. The DR&SS's Agronomy Institute is engaged in crop management research—on-station and on-farm—in maize and other important food crops grown in the highveld and midveld ecologies. In recent years, CBI's budgets for maize breeding research have declined, and private research efforts have increased.

Private Maize Research. There is a growing amount of private maize research in Zimbabwe. The Seed Co-op (formerly called the Zimbabwe Seed Maize Association), which traditionally relied on DR&SS for maize germplasm development, has substantially expanded its own maize breeding work in recent years. The Seed Co-op is especially interested in developing hybrids with better yield dependability for the more marginal production areas, which are farmed mainly by the small-scale freehold and communal producers.

Among the international companies, Pioneer Hi-Bred International began a substantial maize breeding program in 1985 in Harare, while Cargill Seed mainly is screening its materials developed elsewhere for suitability to local maize-growing conditions.

In addition, the Agricultural Research Trust, created in 1980 as the result of the pooling of financial resources of several national agricultural commodity associations, provides facilities and funds to strengthen the agricultural research and extension base for the commercial farmers of Zimbabwe.

International Maize Research. CIMMYT, which established its mid-elevation Africa maize research station at Harare in 1985, has become a valuable source of improved maize germplasm for both public and private maize research organizations. In particular, CIMMYT is providing high-yielding source populations and early generation inbred lines that have strong resistance to maize streak virus.

Research-Extension Linkages

In 1981 the commercial and communal areas extension programs of the government were merged into the Department of Agricultural, Technical, and Extension Services (Agritex) under the Ministry of Agriculture. Since its formation, Agritex has focused on helping small-scale farmers improve agricultural production. Advisory services to the large commercial farmers are mainly provided on an on-call basis.

Private input suppliers are also promoting use of their products in communal areas through village-based demonstrations and sales. The number of private retail outlets has expanded most in those areas where the government has invested in infrastructure and provided production and marketing credits.

Maize Seed Industry

The maize seed industry in Zimbabwe is in private hands although public organizations have played a central role in the development of improved germplasm and hybrids. All of the certified seed sold in Zimbabwe is of hybrid varieties. The Seed Co-op has the sole right to market government-developed hybrids, as well as, of course, its own proprietary hybrids. Seed Co-op hybrid maize seed sales increased from 10,500 tons in 1980 to 28,500 tons in 1989. During this period, seed sales to the communal farms increased from about 40 percent of total hybrid maize sales to 84 percent.

The use of SR 52 has declined in recent years as farmers have switched to more drought-resistant varieties, such as the intermediate maturity (145 days), single-cross hybrid R215 and the early maturity, single-cross hybrid R201 (138 days). In 1989, R215 accounted for 46 percent and R201 for 39 percent of Seed Co-op's annual hybrid maize seed sales. R215 has good resistance to leaf blight and lodging and is marginally higher yielding than R201.

The Seed Co-op has several new hybrids in the marketplace: SC501, a single-cross hybrid that outyields R215 by about 8 percent; ZS225, a single-cross hybrid that outyields R215 by 3 percent but matures and dries down more quickly; and ZS206, a yellow grain, full-season, single-cross hybrid that yields 10 percent more than SR 52.

Maize Production Constraints

Maize yields in the large commercial farming areas exceed 5 t/ha, although there is a "Ten Ton" Club in Zimbabwe with over 200 members who achieve commercial yields of over 10 t/ha. Today, this advanced maize-producing sector is ever-more interested in ways to maintain high yields while lowering production costs.

While yields of smallholders have more than doubled, they are still only one-third those of the large-scale farming sector. Thus, the largest potential for achieving major productivity gains in maize lies in overcoming the various production constraints facing small-scale maize farmers.

Of the fertilizer sold in Zimbabwe, the smallholder sector (communal areas, small-scale commercial farmers, and resettlement areas) accounts for about 30 percent. Fertilizer use, however, is highly skewed in favor of areas with higher rainfall and to higher-value crops. Credit has also been an important factor in fertilizer use by small-scale farmers. The relatively better-off communal farmers, who are located in more-favored production areas, have received production credit, and they regularly apply moderate amounts of fertilizer. The relatively worse-off communal farmers in the less-favored production areas have not received production credit and use little fertilizer. Risk factors such as drought will continue to restrict fertilizer use in many of the communal areas, but there is still considerable scope to increase fertilizer use—and therefore to raise yields—among small-scale farmers.

Most hybrids grown in Zimbabwe were developed for the more-favored production zones. While these hybrids have shown broad adaptation and perform reasonably well, even in the drier areas, there is a need for hybrids that have greater tolerance to drought and various biotic stresses. National maize research organizations, especially the Seed Co-op, are focusing on these germplasm requirements. As more drought-resistant hybrids are developed, maize yields in the marginal zones will improve.

Shortages of labor during peak work periods in the maize cropping cycle frequently occur in the smallholder sector. Draft power can be especially important when labor is in short supply. In areas where the incidence of tsetse fly is not great, animal traction and the moldboard plow have been utilized for over 50 years. More recently, small-scale farmers have been hiring contract tractor services for basic land preparation. Planting, weeding, and harvesting, however, are still done by hand. As small-scale farmers gain greater access to mechanized technology, their maize yields will increase, principally because of better stand establishment and weed control.

Prospects for the Maize Economy

Favorable harvests of maize in 1989 and 1990 in southern Africa were followed in 1991-92 by the worst drought of the 20th century. Maize production in Zimbabwe (and South Africa) declined by 75 percent, and massive imports were needed in 1992 to avert famine. Rainfall was much improved in the 1992-93 season, and Zimbabwe made considerable progress in recovering from this drought and in building up reserve maize stocks again.

Once this recovery has taken place, population growth, rising incomes, intensification of livestock production, and export potential could contribute to future demand growth of 4 percent per year. Population growth alone likely will raise demand at an annual rate of 2.4 percent, and rising incomes will also translate into increasing demand for maize, as human food and especially as a livestock feed.

Although Zimbabwe has been an important exporter of maize in recent years, the prospects that the country will develop a stable long-term export market for maize are unclear. In the past, much of Zimbabwe's maize exports have gone to neighboring countries where civil unrest or discriminatory price policies drastically reduced domestic production, e.g., Mozambique and Zambia. With end of the civil war in Mozambique and as the government of Zambia adopts more favorable price policies, their domestic maize production should increase and their maize import demand should decline.

As an exporter, Zimbabwe must compete with South Africa, which has a comparative advantage as a maize producer (higher yields, more developed marketing channels, ready access to ocean port facilities). Only the large-scale commercial farmers in Zimbabwe can compete on world markets. Their production will depend upon international prices. When world prices are high, the large-scale farmers will plant more maize; when world prices are low, they will plant less. Over time, it is likely that Zimbabwe will increasingly depend on small-scale farmers to supply domestic demand.

Thailand

Thailand's Agriculture

Despite progress in industrialization, agriculture remains a central sector in the Thai economy. It accounts for half of the country's export revenue and provides employment directly to 5 million farm families—65 percent of the population.

Thailand's agriculture is dominated by rice, maize, cassava, and sugarcane production. Rice is grown in all regions and covers 10 million hectares, about one-third of the country's cultivated area. Cassava and maize rank second and third in total area, accounting for slightly more than 3 million hectares. Among the grain legumes, dry beans, soybeans, and groundnuts are important, accounting for about 1 million hectares. Rice and cassava are the chief agricultural exports. In 1990, 4 million

Thailand's agriculture in brief (1991).

Population	57.2 million
Population growth, 1991-2000	1.4%/yr
GDP	US$80,170 million
Per capita GDP	US$1,570
Agriculture vs GDP	12%
Adult literacy	93% total (women: 90%)
Labor force (1989)	56% in agriculture
Total land area	51 million ha
Arable & permanent cropland	20.0 million ha
Cereal crop area	12.2 million ha
Irrigation	4.0 million ha

Sources: World Bank 1993; FAO 1993.

tons of rice, worth US$1 billion, and 7 million tons of cassava chips, worth $600 million, were exported. Maize is the third most important agricultural export, although the quantity exported has declined as demand from the rapidly expanding livestock industry has risen.

The National Maize Economy

Maize was grown on nearly 1.5 million hectares annually during 1990-92, and total production was 3.7 million tons. Thailand's average maize yield, 2.6 t/ha, is relatively high for Southeast Asia, reflecting the widespread use of improved varieties and agronomic practices. Most maize is planted in the central and northern upland areas adjacent to lowland rice lands. Soils in these areas are generally neutral to slightly alkaline. Maize is produced in the rainy season over a 4-month growing cycle. Planting occurs in April-May and harvesting in July-August.

Thailand's maize economy in brief (1990-92).

Maize area		1.45 million ha
Maize yield,		2.6 t/ha
Maize production		3.71 million t
Per capita utilization		53 kg/yr
Net maize imports		−722,000 t/yr
Maize grain color		0% white
Maize used as	human food	1%
	livestock feed	95%
Area planted to	unimproved varieties	0%
	improved OPVs	75%
	hybrids	25%

Source: CIMMYT 1994.

Thais consume little maize directly, other than a small amount of roasted or boiled green ears. Most maize is either exported or fed to livestock. Pacific Rim nations have been the major export customers. In 1985, 3 million tons of maize, 75 percent of national production, were exported, but maize exports dropped to 1.2 million tons by 1989-90 and to 0.7 million tons by 1990-92. Thailand is one of the "Asian tigers," and its prosperous consumers can now afford more meat in their diets. Large intensive livestock-producing facilities have been established to meet this demand. Thailand has also become an major exporter of meat and eggs to Pacific Rim countries.

Feeding maize to domestic livestock and poultry—and then exporting these higher-value products—has been economically beneficial for Thai agriculture. Industrialization of maize also came at a fortuitous time because some maize exported from Thailand was having aflatoxin problems. In Thailand, maize matures during the rainy season, making it difficult to dry the grain sufficiently before storage. Grain molds that produce aflatoxin thrive in wet grain. However, grain fed to local livestock and poultry is stored for a short period, reducing the opportunities for aflatoxin to develop.

Principal Maize Ecologies

Thailand's main maize-growing areas are in the central, northern, and northeastern regions of the country (Fig. 10.3). The central plains account for 1.0 million hectares, the northern region for 0.4 million hectares, and the northeastern region for 0.3 million hectares. The principal soil groups are reddish brown laterites and black soil types.

Maize is grown almost entirely as a rainfed crop in Thailand. A bimodal rainfall distribution prevails in the major maize-growing areas. Early rains start in mid-April and the main rainy season begins in mid-May. The main maize-growing season starts during April-May. The high humidity and temperatures during the harvesting period may cause aflatoxin problems in stored grain. The secondary maize-growing season starts in July-August, with harvest coming during a relatively drier period. Eighty percent of the maize area is planted in the main growing season, with farmers producing other field crops such as sorghum, mungbean, and groundnut as the second crop. Drought affects 10 to 20 percent of the total maize area.

Over 30 diseases attack maize in Thailand. But only six cause serious losses, if not properly controlled: downy mildew (primarily caused by *Peronosclerospora sorghi*), charcoal and anthracnose stalk rots (caused by *Marcophomina phaseolina* and *Collectotrichum graminicola*), southern rust, and aflatoxin, caused mainly by Aspergillus flavus (Table 10.4).

The Maize Research System

After World War II, Thailand followed a policy of diversifying food production to reduce dependence on rice. Malaria eradication programs in the 1950s and 1960s allowed the habitation of the central and northern upland areas. Forests were cleared, and efforts to develop maize as a commercial crop were launched in these new areas. To raise yields, the Thai Department of Agriculture (DOA) released the open-pollinated variety, Guatemala (Tiquisate Golden Yellow) in 1951. The variety was selected from Mexican-Caribbean-Central American germplasm supplied by the Rockefeller Foundation-Mexican maize program. In addition the maize breeders of the DOA raised the yield of Guatemala using controlled mass selection techniques, leading to the release of Phra Bhutthapbat (PB 1-12), which was extensively cultivated by farmers.

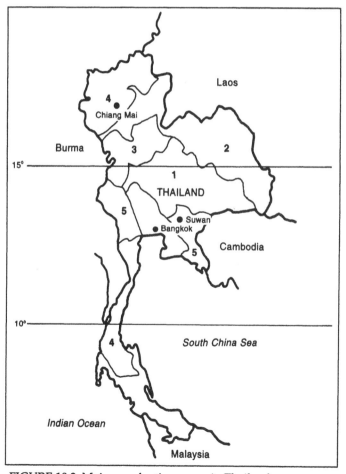

FIGURE 10.3 Maize production zones in Thailand.

TABLE 10.4 Characteristics of maize production zones in Thailand.

	Production zone				
	1	2	3	4	5
Area, 000 ha	1,375	286	350	97	122
Ecology	tropical	tropical	tropical	tropical	tropical
Grain type	yellow flint	yellow flint	yellow flint	yellow flint	yellow flint
Growing season	major and minor	major	major	major	major
Maturity	late	late	late	late	late
Moisture stress	sometime	often	sometime	rare	often
Soil type	normal	normal	normal	normal	normal
Biotic stresses [a]					
Downy mildew	●●●●●	●●●	●●●	●●	●
Southern leaf blight	●●	●●	●●	●●	●●
Southern rust	●●	●●	●●	●	●●
Stalk rot	●●	●●	●●	●●	●●
Ear rot	●●	●●	●●	●	●●
Thrips	●●	●●	●	●	●
Locust	●●	●●	●	●	●
Oriental corn borer	●●	●●	●●	●	●
Fall armyworm	●	●	●	●	●

[a] Stress ratings: O Biotic stress not present in region. ● Present but not of economic importance. ●● Some economic losses. ●●● Significant economic losses. ●●●● Severe economic losses. ●●●●● Maize cannot be grown unless a resistant variety is grown or chemical control is applied.
Source: CIMMYT.

In 1960, the Rockefeller Foundation established a collaborative maize research program with Kasetsart University. At the time, Thailand grew less than 200,000 hectares of maize. The Corn Breeding Project of Kasetsart University, a collaborative effort of the DOA and the Rockefeller Foundation, had a major impact on maize research and development in Thailand, other Southeast Asian countries, and later in other lowland tropical areas of the developing world. The Rockefeller Foundation provided technical assistance, new sources of useful exotic germplasm, and provided scholarships for local scientists. The foundation also was heavily involved in the overall development of Kasetsart University. The Rockefeller Foundation's Inter-Asian Corn Program, also headquartered in Bangkok, brought in tropical germplasm during the 1960s from Mexico, Central America, the Caribbean, and Brazil.

Today, maize research in Thailand is carried out by both public and private organizations. In the public sector, the staffs of Kasetsart University and the DOA are involved in maize research. In germplasm improvement, Kasetsart University devotes 60 percent of its research effort to population improvement and 40 percent to inbred-line development. The DOA concentrates on population improvement and open-pollinated variety development. The private sector, in contrast, focuses on developing proprietary hybrids. Much of this private research, however, depends heavily on public-sector maize germplasm (Suwan-1, other populations, inbred lines) developed primarily by scientists at Kasetsart University and CIMMYT.

The Suwan germplasm complexes (Suwan-1, Suwan-2, and Suwan-3), which are resistant to downy mildew are an exceptional maize research achievement. These populations were originally developed by crossing Thai Composite #1—a synthetic with 36 varieties from Mexico, Central America, and the Caribbean—with several germplasm sources of downy mildew resistance from the Philippines (Table 10.5). In 1987,

TABLE 10.5 Outstanding maize varieties and hybrids released by public-sector organizations in Thailand.

Name	Type[a]	Grain color	Year of release	Germplasm origin
Suwan-1	OPV	Yellow	1974	Thai Composite #1 & TDMR
Suwan-2	OPV	Yellow	1976	Early maturing fraction of Suwan-1
Suwan-3	OPV	Yellow	1988	Suwan-1
Nakhornasawan	OPV	Yellow	1989	CIMMYT Pop. 28 x Suwan-1
KSX 2301	SC	Yellow	1983	Ki3 x Ki11 (Suwan)
KTX 2602	TWC	Yellow	1984	Ki3 x Ki11 x Ki 20
Suwan 3101	TWC	Yellow	1991	Ki 27 x Ki 28, Ki 21

[a] OPV = open-pollinated variety; SC = single-cross hybrid; TWC = three-way cross hybrid.
Source: CIMMYT.

Suwan-1 (cycle 9 of improvement) covered 85 percent of the total maize area. Suwan-2 was grown on about 5 percent of the area, mainly localities requiring early maturing varieties for drought escape and where baby corn is produced (for export to Japan and other Pacific Rim countries).

In 1989, the DOA maize breeding program released the open-pollinated variety Nakhornasawan, which was developed by crossing CIMMYT Population 28 with Suwan-1. This variety is higher yielding than Suwan-1 and has excellent resistance to downy mildew. It has proved popular with farmers—it is now estimated to be grown on 30 percent of the total maize area in Thailand.

CIMMYT collaborated with the Thai national maize program during the 1980s to incorporate downy mildew resistance into three of its most outstanding lowland tropical maize populations: 22, 28, and 31. Populations 22 and 28 are full-season materials and population 31 has a maturity similar to Suwan-2.

Maize Seed Industry

The production of high quality maize seed in Thailand was initiated by the government in 1977 through the Seed Development Project, under the responsibility of the Seed Division of the Department of Agricultural Extension, which receives basic or foundation seed from the DOA and Kasetsart University. The foundation seed is multiplied under contract with seed growers, dried, processed, tested for germination, packaged, and then distributed to farmers.

About 75 percent of the maize seed produced comes from the planting initiated in the second (late) season, with the remainder from the first (early season). Although higher yields are obtained from first-season planting, grain quality is generally lower due to the high humidity and temperature during and following the harvest.

From 2,000 to 2,500 tons of maize seed (virtually all open-pollinated varieties) is produced annually by the Seed Division, which represents a quarter of Thailand's maize seed production. Private growers produce the other 75 percent.

After the development and release of Suwan-1 in 1974, the government promoted private participation in the maize seed industry. Tax incentives and exemptions (imported equipment) were established to encourage private companies—domestic and foreign—to produce and distribute high quality maize seed, either from public varieties or proprietary hybrids. By 1990, there were nine private maize seed firms, which had processing plant capacities ranging from 1,000 to 3,000 tons per year (Table 10.6).

Maize Production Constraints

Initially, Thailand's maize production strategy centered on expanding the maize area. More recently, however, higher yields have been stressed. In Asia, Thailand has among the highest proportions of farmers using improved maize germplasm and among the lowest proportion using fertilizer. Some 75 percent of all farmers plant improved open-pollinated varieties and another 25 percent plant hybrids. Agronomic research data have consistently shown that the application of nitrogen fertilizer is profitable for farmers. However, the profitability of using phosphorus and potassium depends greatly on the soil groups in question. Yet about half the farmers do not apply fertilizer to maize. Among those who use fertilizer, most apply sub-optimal dosages. Harrington et al. (1991) attributed the low adoption of fertilizers in Thailand to their high cost relative to the price of maize, to the promotion of compound fertilizers with nutrient formulas inappropriate for local soils—resulting farmers paying for nutrients for which there is no yield response—and to government policies that discouraged the import and use of low-cost, high-analysis fertilizer formulas like urea or diammonium phosphate.

Other maize yield constraints are the use of slightly lower yielding seed than the best available, planting at suboptimal densities, and ineffective and costly weed control, done mainly by hand.

Prospects for the Maize Economy

The future of the maize economy in Thailand is highly dependent upon prices in the international market and upon rates of growth in

TABLE 10.6 Principal maize seed enterprises in Thailand.

Organization	Seed products	
	OPVs	Hybrids
Public organizations		
Seed Division, Department of Agricultural Extension	●	
National Corn and Sorghum Research Center (Suwan Farm)	●	●
Department of Agriculture	●	
Private companies		
Bangkok Seed Industry Co. Ltd	●	●
Cargill Seed, Ltd.	●	●
Charoen Seeds, Ltd. (DeKalb-Pfizer Genetics)	●	
Charoen Pokphand Agri-Industry, Ltd	●	●
Pacific Seed, Ltd.		
Pioneer Hi-bred International, Ltd.	●	
Super Seeds, Ltd.	●	
Thai Seed Industry, Ltd.	●	●
Ciba-Geigy. Ltd	●	

Source: De Leon et al. 1988.

domestic meat consumption. Export of maize will continue to decline as rising incomes in Thailand enable people to eat more maize-fed livestock and poultry and as opportunities for exporting meat and eggs expand.

The government policy to promote a private seed industry will result in a growing demand for maize hybrids. Until now few hybrids have achieved a 20 percent yield superiority over the open-pollinated maize variety, Suwan-1. However, maize breeders at Kasetsart University and in private seed companies are developing outstanding inbred lines based upon Suwan-1 germplasm and other tropical materials. The new single-cross and three-way cross hybrids that are becoming available are sufficiently superior in yield potential to Suwan-1 for farmers to be persuaded to adopt them.

Significant growth in average national maize yields should be expected in Thailand during the 1990s. The introduction of high-yielding hybrids will serve as a catalyst for increased fertilizer use and improved land preparation through mechanization and better weed and pest control through integrated pest management.

China

China's Agriculture

China's population of 1.15 billion is growing at 1.3 percent per year, adding about 15 million people annually. Approximately 700 million are engaged directly in agriculture, producing about 27 percent of the country's GDP. Varied environments ranging from temperate to tropical allow China to produce a wide range of crops and lend stability to production. The major grain crops are rice, wheat, maize, millet, sorghum, and barley. The major grain legumes are soybeans, groundnuts, dry beans, and peas. The major tuber crops are sweet potatoes and potatoes.

China's land area is about 933 million hectares, of which nearly 97 million hectares are currently cultivated or in permanent crops. Cereal crops are grown on about 91 million hectares. Nearly half of the arable land is irrigated and more than half of this area grows at least two crops per year. This intensive production system allows China to meet the basic food needs of its population, which constitutes 21% of the world's total populations, on less than 8 percent of the world's arable land. The small proportion of land that is arable (10% as compared with 75% in India) has forced China to rely heavily on science and technology to raise yields to feed its enormous population.

During the 1980s, China's agricultural sector performed exceptionally well. The agricultural economy grew over 6.1 percent per year, outper-

China's agriculture in brief (1991).

Population	1.15 billion
Population growth, 1991-2000	1.3%/yr
GDP	US$369,651 million
Per capita GDP	US$370
Agriculture vs GDP	27%
Adult literacy	73% total (women: 62%)
Labor force (1990)	60% in agriculture
Total land area	933 million ha
Arable & permanent cropland	97 million ha
Cereal crop area	91 million ha
Irrigation	45 million ha

Sources: World Bank 1993; FAO 1993.

forming all other countries during the decade. Today, its cereal crop yields are the highest in Asia except for Japan, South Korea, and Taiwan. China is one of the world's largest net food importers, even though imports account for only 5 percent of total supply. In recent years, China also has become an important exporter of maize, particularly to Japan, and of grain-derived products such as live animals and meat.

China has high per capita direct grain consumption. Only about 10 percent of the protein in Chinese diets is from animal sources, as compared with 35 percent worldwide and 21 percent in developing countries. Nevertheless, the average diet exceeds WHO nutritional standards and is markedly better than diets in other low-income countries. While some regional differences in per capita food availability exist, China's food procurement and distribution system has succeeded in providing basic food nutrients to most of the population most of the time.

During the 1980s, to improve incentives and management of the agricultural economy, the Chinese government restructured rural institutions, shifting control over resources from collectives to individuals (Haldore Hanson, personal communication). A major introduction has been the production responsibility system (PRS), the generic name for contracting arrangements that define the rights and responsibilities of owners of assets (state, collective, or private), on the one hand, and of managers of these assets, on the other. The PRS variant used in agriculture is the *bao gan dao hu* (BGDH) or "contracting all actions to the household" system. Under BGDH, the individual household has replaced the collective (production team) as the basic farm management and production unit. At higher levels, the commune has been renamed township (*xiang*), and its economic role has been restricted to managing commune and brigade enterprise development, while the brigade is

China's maize economy in brief (1990-92).

Maize area		21.4 million ha
Maize yield,	·	4.5 t/ha
Maize production		97.2 million t
Per capita utilization		83 kg/year
Net maize imports		−1.7 million t/yr
Maize grain color		20% white
Maize used as	human food	33%
	livestock feed	57%
Area planted to	unimproved varieties	3%
	improved OPVs	7%
	hybrids	90%

Source: CIMMYT 1994.

now named village (*cun*) and is losing much of its managerial and technical staff. Management of collectively owned land is contracted to households, usually in proportion to household size or labor force. The household is obligated to pay taxes, contribute to collective welfare funds, provide its share of state procurement requirements, and offer labor to maintain or construct public infrastructure. The household retains all remaining output. BGDH has given substantial rewards for increased production.

The National Maize Economy

China has nearly 500 years of experience with maize. Portuguese explorers brought the first maize seed to China. According to a county record published in 1511, maize was first grown along the Fukien coast in southeastern China around 25° N latitude. That was only 19 years after Christopher Columbus first landed on Hispanola in the Caribbean. Subsequently, maize spread fairly quickly, both northward and westward. Within 200 years, maize cultivation had become well-established in China. During the past 50 years, the national maize area has doubled, while production, propelled by the introduction of high-yielding hybrids and increased fertilizer use, has increased eight-fold.

Maize is the most important cereal in China after rice and wheat. China is the largest maize producer in the developing world. During 1990-92, about 97 million tons of maize grain were produced annually on 21.4 million hectares, with national yields averaging 4.5 t/ha.

Approximately 70 percent of the maize has a yellow flint-dent grain type, and 30 percent has a white dent-flint grain type. About 60 percent of the maize crop is utilized for human consumption and 30 percent for livestock feed. Improved genotypes are planted on about 92 percent of the total maize area (90% to hybrids and 7% to improved open-pollinated varieties). In provinces that have a high concentration of maize (Shandong, Liaoning, and Jilin), virtually all of the maize area is planted to hybrids, with single crosses being the principal type.

Principal Maize Ecologies

Nearly two-thirds of China's maize area is located within temperate environments. Most of the rest is subtropical to tropical. Six broad production zones can be delineated (Table 10.7). In general terms, maize-growing areas can be divided into spring- and summer-growing areas.

Maize is the most important crop on the fertile soils of the temperate northeast region (Fig. 10.4). This region—located north of 40° N latitude—corresponds to the "spring maize" zone. Maize is sown in April and is the only crop planted each year. Maize is the primary crop in Jilin and Liaoning provinces, where it is grown on 60 to 65 percent of the cultivated land (Jilin is the largest maize-exporting province in the country). Being similar to the U.S. Corn Belt in climate and soil types, the spring maize region attains yields of 5.5 to 6.5 t/ha. Yield losses due to stalk rots, leaf spot, and northern leaf blight are serious. Head smut is a problem in cold, dry areas where rotations are not practiced. A few other diseases, such as maize dwarf mosaic virus, maize rough mosaic virus, and kernel rot also cause damage in some areas.

Maize is also grown as a spring-planted monocrop in the temperate northwest region. Northern leaf blight, maize dwarf mosaic virus, ear rots, and borers are the main biotic stresses. Soil toxicities and moisture stresses are problems in some areas. Overall, however, yields are high.

The temperate north China plains region accounts for 3.6 million hectares, or 18 percent of the total national maize area. About 70 percent of the maize fields are double-cropped with winter wheat, using a relay cropping system in which maize is planted between the rows 3 to 4 weeks before the winter wheat is harvested. Maize yields in this region range from 5.5 to 6.0 t/ha. Stalk and kernel rots and northern and southern leaf blights can cause serious economic losses.

Some 4.9 million hectares of maize are grown in the temperate central region. In this mid-latitude climate, the summer season lasts a little longer than it does in higher latitudes. Maize is planted after winter wheat in a sequential cropping system. In the surrounding mountainous areas, maize is planted as a spring crop after winter wheat. The yield potential is 5 to 6 t/ha.

In the southeastern region, maize is mostly grown as a spring and summer intercrop in multiple-cropping systems. There is also a small

TABLE 10.7 Characteristics of maize production zones in China.

	Production zones					
	1	2	3	4	5	6
Area, million ha	5.7	3.6	1.0	0.7	4.9	4.6
Environment	temperate	temperate	temperate	temperate/subtropical	temperate	subtropical/tropical
Grain type	yellow dent, yellow QPM	yellow dent, yellow QPM	yellow dent	mixed colors	white QPM	yellow dent, mixed colors (dent)
Growing season	spring	spring, summer	spring, summer	spring, summer	spring	spring, summer, autumn
Maturity	early	early, int.	early, int.	early, int., late	early	early, int.
Moisture stress	sometimes	rare	sometimes	rare, sometimes	sometimes	rare
Soil type	normal	normal	normal	acid	acid	acid, normal
Biotic stresses[a]						
Northern leaf blight	●●	●●	●●	●●	● 1/2	●
Puccinia spp.	●	●	●	●	O	●
Southern leaf blight	O	O	●	●	O	●●
Maize dwarf mosaic	●	●	O	O	O	O
Maize rough mosaic	O	O	O	O	●●	O
Head smut	●	●	●	O	O	O
Ear rot	●	●	●	●●	●●	●●
Earworm	O	●	●●	●	O	O
Borer	●●	●●	●●	●	●	●

[a]Stress ratings: O Biotic stress not present in region. ● Present but not of economic importance. ●● Some economic losses. ●●● Significant economic losses. ●●●●Severe economic losses. ●●●●● Maize cannot be grown unless a resistant variety is grown or chemical control is applied.

Source: CIMMYT.

amount of autumn maize grown in Zhejiang province. Maize is first sown in nursery plots and then transplanted to paddy fields in July or August after the harvest of early maturing rice.

In the southwest region, maize is grown as part of multiple-cropping systems, either as a spring or summer inter-crop, at elevations ranging from 250 to 3,000 meters above sea level. Because of the low fertility of the red soils and the many rainy or foggy days that reduce solar radiation, the yield potential of maize in this region is much lower than in the northern regions. Average yields are in the 3.5 to 4.5 t/ha range. Northern leaf blight, stalk and ear rots, and head smuts in highland areas and southern leaf blight in subtropical and tropical areas are diseases of economic consequence. In addition, various stalk borers, rootworms, and cutworms often cause damage.

The Maize Research System

Maize improvement research in China began in the 1930s. Plant breeders developed several double-cross hybrids of a flint type, however none were released to farmers because the yields were not exceptional, and under the Kuomintang regime no government agencies existed for seed production. After the creation of the People's Republic of China in 1949, maize improvement research was resumed. The last 40 years of maize research and technology development in China can be

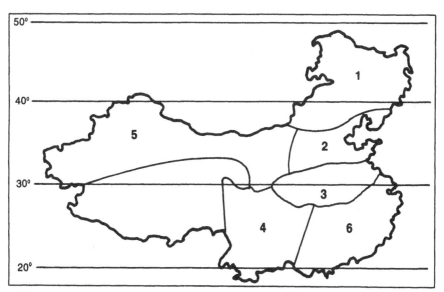

FIGURE 10.4 Maize production zones in China. (1) Northeastern region. (2) Northern region. (3) Central Region and Yellow River. (4) Southwest region. (5) Northwest region. (6) Southeast region.

divided into four phases. During the 1950s and early 1960s, many crops research institutes were established and the commune administration assumed the role of agricultural extension. In the second phase, 1966-76, the Cultural Revolution was a period of disruption, destruction of equipment and publications, and closing of academies and research institutes, resulting in serious setbacks in maize research and development. The third phase, 1976-80, served to restore the previous research structure. The fourth phase, which continues today, has been characterized by a revival of scientific and technological development.

Maize research is carried out by Chinese Academy of Agricultural Sciences (CAAS) and 10 provincial academies of science. In addition, some of the prefectural institutes have been active in maize breeding and have made important contributions. The Institute of Crop Breeding and Cultivation in Beijing is the largest maize improvement effort, with research operations under way at a number of its institutes. Especially strong provincial maize research programs are carried out at the maize institutes of the Shandong Academy of Agricultural Sciences in Jinan and the Jilin Academy of Agricultural Sciences in Gongzhuling, 900 kilometers north of Beijing. In the southwest region, the Maize Research Institute of the Guangxi Academy of Agricultural Sciences in Nanning and the Yunnan Academy of Agricultural Sciences Food Crops Research Institute in Kunming have active maize improvement and crop management research programs.

Maize Research Achievements

Hybrids based on U.S. dent germplasm are used on 70 percent of China's maize-growing area. During the 1950s, hundreds of inbred lines were isolated from Golden Queen, the U.S. dent variety brought to Shanxi in 1936. Because many of these selections suffered from lodging and susceptibility to northern leaf blight, they had to be crossed with local sources of resistance. The original double-cross hybrids introduced during the 1950s and 1960s were based on elite Chinese flint germplasm crossed with U.S. Corn Belt lines, including 38-11, L289, W19, W20, W24, and M14. These double-cross hybrids—the most popular being Nungda 4 and 7—were well received by Chinese farmers in the 1960s, who appreciated their higher yield potential, short stature, and plant uniformity. By 1965 double-cross hybrids covered 330,000 hectares in Shanxi province. However, in 1966 an epidemic of northern leaf blight in North China badly damaged most hybrids. This epidemic discredited the hybrids in the eyes of many farmers. The Cultural Revolution, launched in 1966, sent many researchers back to the land, where they were assigned to work in various communes. With access to interna-

TABLE 10.8 Area planted to popular hybrids in
China, 1982 and 1986.

Name	Developed by	Area (000 ha) 1982	1986
Zhongdan 2	CAAS	1,602	2,078
Danyu 6	Liaoning	1,082	42
Zhengdan 2	Henan	834	154
Luyuandan 4	Shandong	696	164
Jidan 191	Jilin	490	660
Sidan 8	Jilin	180	1,039
Yendan 14	Shandong	44	986
Danyu 13	Liaoning	–	707
Luyu 2	Shandong	–	676
Total		4,928	6,506

Source: De Leon et al. 1988.

tional germplasm sources effectively cut off, Chinese breeders had to turn to local varieties in search of resistance to northern leaf blight. By the end of the 1960s, a new group of blight-resistant single-cross hybrids had been developed for use in North and Northeast China.

During the 1970s, Chinese breeders again had access to new inbred lines from public institutions in the United States and other countries. Inbred lines, such as Mo17, Oh43, A619Ht, B73, Bup44, and Egypt 205 were used extensively. B73, the outstanding Iowa State University inbred, however, has not been used since it was found to be susceptible to southern leaf blight. In 1980, 37 percent of the hybrids grown in China were crosses of local and foreign inbreds (Li 1988). The most popular of the new hybrids has been Zhongdan 2, which is broadly adapted and has good resistance to leaf blights and head smut. It has been grown on as many as 2 million hectares in a single season (Table 10.8).

The development of inbreds has been hampered by the narrow spectrum of genetic variability from which Chinese maize breeders select. An elite inbred line developed in one institution is shared with other maize research programs and often becomes a common component in a series of hybrids bearing different numbers. As an example, there are at least 17 single-cross hybrids released in China that have an inbred pedigree of Yellow Early 4 in common, thus creating the risk of genetic vulnerability (Li 1988).

Since 1974, Chinese maize researchers have also been breeding for improved protein quality in maize, and they expect high-lysine and high-oil maize will be utilized for livestock feeding and industrial purposes. A QPM open-pollinated variety, Tuxpeño 102, has been released along with several hybrids. The QPM hybrid Lu Dan 203 yields within 4 percent of the normal hybrid, Zhongdan 2. Luyu 13, a QPM hybrid released in 1991 in Shandong province, was planted on 100,000 hectares by 1993. It won first prize at the Second Exhibition of Invention and Technology of Shandong province in 1993. Specialty hybrids have also been developed. The waxy hybrid, Lu Nuo 1, was developed for green ear consumption. Development of sugary-2 hybrids is also under way.

Several single-cross hybrids, Liu Yu 10 and Liu Yu 11, which have been recently developed in Shandong, show 5 to 8 percent higher yield potential than Zhongdan 2. The newly released hybrids J159 and J180 yield even more.

Chinese maize breeders in the southwestern and southeastern maize regions have made use of CIMMYT's tropical germplasm to release several improved open-pollinated varieties, Mexican White and Mexican White 94, based on Tuxpeño germplasm (Tuxpeño-1 and Tuxpeño Planta Baja C_{15}). These two varieties are grown on about 100,000 hectares in Guangxi province.

Research-Extension Linkages

With 180 million farm families engaged in agricultural production, organizing China's technology delivery systems to provide continuing education to the nation's farmers is a demanding task. It became more difficult when the communes were abolished in the early 1970s. The sharply diminished roles for brigades and teams in production decisions have made necessary a new approach to extension. In 1982, a decision was taken to develop the system around county governments.

Extension operates as an independent system under the Ministry of Agriculture. The National Technical Extension Center is primarily responsible for overall planning at the national level. At the provincial and prefectural levels, bureaus of agriculture are responsible for coordinating extension activities. Below the prefectural level, county research institutes have been transformed into agro-technical extension centers (ATECs), which combine the formerly separate activities of extension, research, plant protection, seed production, fertilizer distribution and soil analysis, farm machinery, and animal husbandry. These ATECs act as focal point for extension work in the country. About 300 ATECs existed in 1985. By 2000 the government expects to have established ATECs in all 2,300 counties in the country. Each ATEC, in turn, is supported by a network of township agro-technical extension stations and village extension groups who assist farmers. Actual extension contacts with the farmer are primarily carried out by farmer-technicians who make farm visits and conduct farm demonstrations.

Provincial and prefectural research institutes and county ATECs carry out applied maize research and field testing programs. The ATECs also provide soil testing and crop protection services, supply seed and fertilizer, and provide training and technical support to township extension staff and farmer-technicians.

Maize Seed Industry

China's maize seed industry is entirely a public-sector activity operated as a not-for-profit enterprise. Initially, the people's communes and brigades produced maize seed (primarily hybrids) under the supervision of the county government. Foundation seed was increased on county seed farms or in a few communes qualified to do such work. Commercial hybrid seed was generally produced in brigades assigned by country authorities and by contract. The seed departments above the county level, administered by the prefectural and provincial governments, were responsible for policy making, planning, and coordination related to hybrid maize production.

In 1974, China established the National Seed Corporation (NSC). There are about 2,300 NSC-affiliated, but locally financed and administered, seed companies that handle multiplication, processing, storage, certification, and distribution of certified seed. Breeder's seed is supplied primarily by provincial agricultural research institutes. Multiplication of foundation and certified seed is carried out by specialized seed farms managed by, or under special contract to, the local seed companies. These seed companies contract individual farmers to produce certified seed. The system of contract growing with many small-scale landholders in a contiguous block has been difficult to manage and supervise. The central General Bureau of State Farms also operates a parallel system of seed companies and contract growers.

Although the basic mechanisms are in place, processing and quality assurance procedures are inadequate. Most commercial seed is dried in the sun. In the northeast, however, where the temperature drops quickly in the fall, the seed moisture content remains too high. As a result, farmers often get maize seed that has low germination percentages, which requires that they plant at higher seeding rates, thus adversely affecting yields and raising production costs.

Seed quality has improved somewhat since the NSC was established. Even so, the seed sold in the market is not graded or certified. The government's current program of seed industry development is aimed at modernizing over 300 seed production, processing, and distribution centers by 2000. This process has begun in the Northeast.

Some transnational private maize seed companies have established collaborative networks with China for testing private hybrids from the U.S. Corn Belt and Europe in the temperate maize-growing areas. These foreign hybrids have yield advantages of 10 to 15 percent over the best

locally available hybrids and are superior in plant type and other genotypic traits. So far, however, these foreign hybrids are not being produced commercially due to the lack of suitable agreements on intellectual property rights that protect private hybrids.

Maize Production Constraints

Chinese farmers have made great strides in improving productivity
in maize cultivation, but various factors constrain maize yields. The
most important are improper fertilization practices, poor quality seed,
and the use of hybrids with inadequate genetic yield potential.

Fertilizer Supply and Use. Although farmers in China apply considerable amounts of fertilizer to their maize crops, soil fertility management
remains an important constraint to increased productivity. Nitrogen fertilizers have been readily available because 90 percent of the demand is
met through domestic production. In contrast, only 72 percent of phosphorus fertilizers and 6 percent of potassium fertilizers are manufactured domestically, so they have been in short supply. Farmers, as a result, infrequently apply balanced fertilizer nutrient formulations to
maize, leading to inefficient use of the fertilizer nutrients they do apply.

In addition, fertilizer supplies often arrive too late in the season to allow farmers to make timely application, and farmers complain about
the quality of the local fertilizer manufactured in small-scale factories.
Improvements in product quality and distribution systems are needed
to achieve higher yields.

Also, as cropping systems intensify, more precise monitoring of nutrient balances is required. Deficiencies of secondary nutrients and mi-
cronutrients are affecting the crop's ability to make optimum use of the
primary nutrients. Better soil testing and analysis systems are needed to
monitor and amend soil nutrient balances, and extension efforts to upgrade the soil fertility management skills of Chinese farmers will have
to increase.

Seed Production. Increases in maize yield and production are retarded by the poor quality and inadequate supplies of breeder and
foundation seed, by lax supervision of the many small-scale contract
growers involved in producing seed, and by poor seed processing and
storage systems. Consequently, hybrid seed sold to farmers often has
low germination, is poorly cleaned and graded, and lacks appropriate
seed treatments such insecticides or fungicides.

The government plans to achieve maize production targets by developing new hybrids that have higher levels of tolerance to diseases, insects, and other stresses; by upgrading seed production systems in order to supply sufficient high quality seed of the latest hybrids to plant

90 percent of the maize area; and by strengthening extension systems to help farmers improve their crop management practices, especially fertilizer use and soil fertility management.

The government has resisted allowing private maize seed companies to operate. Differences of opinion exist on control over, and multiplication of, breeders' seed, pricing, and repatriation of profits. If these issues were settled and private companies were given the green light to operate aggressively, private hybrids would probably spread rapidly on Chinese farms.

Germplasm Development. The development of higher yielding hybrids with improved resistance to abiotic and biotic stress is held back by the lack of breeding material with a broad genetic base. China needs access to elite breeding materials that have greater disease and insect resistance. China's strategy is to obtain superior breeding material from countries possessing similar production environments.

Worldwide, the most elite germplasm for temperate areas is held by private seed companies that operate extensive germplasm development programs. Chinese maize scientists cannot access this germplasm freely, unlike germplasm from public-sector maize breeding organizations, which have liberally shared their elite germplasm.

Institutional Coordination. The organizational linkages within and among national research institutes concerned with maize research need to be greatly improved in China. Inadequate communication, visits, and coordination among national and provincial academies, research institutions, seed companies, and extension agencies hamper technology transfer and foster duplication of efforts.

The expansion of ATECs promises to improve the coordination of agricultural support services to the farmer. The focus on county and township operational centers and the close integration of extension, input-supply, and crop protection services should be particularly beneficial to the increasingly complex production systems that are emerging in Chinese agriculture. The consolidation of services also should improve the flow of information between the research institutes and agricultural colleges on the one hand and the millions of small-scale farmers on the other.

Prospects for the Maize Economy

Maize will become increasingly important as a livestock feed and for industrial processing into starch and modified starches. The production target of the Eighth 5-Year Plan was to reach 90 million tons by 1995. However by the time the 5-Year Plan was launched, average production already had exceeded 90 million tons. It is likely that national maize production will soon be above 100 million tons. Increased production has been obtained almost entirely through yield gains. Indeed, the total area planted to maize is expected to shrink slowly. Improved soil fertility management, higher yielding hybrids, and better quality inputs, especially seed and fertilizer, will be the primary sources of growth in maize yields.

Guatemala

Guatemala's population of 9.5 million people is growing at 2.9 percent per year. Most of the population lives in highland areas. In the higher-elevation western and central highlands, the people are mainly Amerindians who speak one of several dozen Mayan languages and dialects. In the lower-elevation eastern highlands, Guatemala City, and the lowland area along the Pacific coast, most of the population is primarily Ladino, a cultural definition referring to those indigenous people who follow "western" cultural practices and speak Spanish as their first language.

Guatemala's Agriculture

Agriculture is the mainstay of the Guatemalan economy. Half the labor force is engaged in agriculture, which has a dualistic economic orientation. Guatemala's agricultural exports—coffee, sugar, cotton, vegetables, flowers—account for most of the foreign exchange earnings. The major commercial agricultural area is found along the Pacific coastal plain. This strip, 35- to 60-kilometer-wide strip is the center of cattle, cotton, and sugarcane production, as well as of basic food crops

Guatemala's agriculture in brief (1991).	
Population	9.5 million
Population growth, 1991-2000	2.9%/yr
GDP	US$9,353 million
Per capita GDP	US$930
Agriculture vs GDP	26%
Adult literacy	55% total (women: 47%)
Labor force (1989)	49% in agriculture
Total land area	11 million ha
Arable & permanent cropland	1.9 million ha
Cereal crop area	0.8 million ha
Irrigation	77,000 ha

Sources: World Bank 1993; FAO 1993.

Guatemala's maize economy in brief (1990-92).

Maize area	0.65 million ha
Maize yield,	1.9 t/ha
Maize production	1.23 million t
Per capita utilization	143 kg/yr
Net maize imports	126,000 t/yr
Maize grain color	85% white
Maize used as human food	72%
livestock feed	22%
Area planted to unimproved varieties	31%
improved OPVs	48%
hybrids	12%

Source: CIMMYT 1994.

production. Most farms are intermediate in size. In the piedmont areas, considerable coffee is produced in addition to basic food crops.

Food production in much of the central and western highland is of a subsistence nature. Farmers' landholdings usually are small (under 2 ha) and fragmented. Wheat and vegetables are the principal crops. Maize is primarily a subsistence crop. In the lower-elevation eastern highlands, agriculture is more commercially oriented. Most farmers are small landholders engaged in basic food production. New government land settlement projects are opening the piedmont and lowland areas in the northern portion of the country to commercial agriculture.

The National Maize Economy

Botanists consider the lowland area of what today are southern Mexico and northern Guatemala to be the center of origin of domesticated maize. Thus, maize has been the staff of life of the Guatemalan people for thousands of years. So important was maize to the Mayan-Quiche culture that the great Mayan epic poem, the Popo Vuh, portrays maize as the mother of humankind.

Maize remains the staple food for most Guatemalans. Virtually every part of the plant is used. The grain is the principal energy source in most diets. The husk is used to wrap maize dough that is steamed to make tamales. During the rainy season, the lower leaves of the plant are stripped off to serve as green fodder for livestock. During the dry season, the plant stalks provide forage for animals. The cob is used as a cooking fuel. The remaining parts of the plant are used as bedding in farm stables. Subsequently the mixture of this bedding material, animal urine, and manure becomes organic fertilizer that is applied to the maize fields at the next planting.

About 600,000 hectares are planted to maize. That is 92 percent of the national cereal crop area and 40 percent of the total arable land and permanent cropland base. National production during 1990-92 averaged 1.23 million tons. It grew by 3.8 percent per year between 1960 and 1990. Average yields over this 30-year period have risen from 0.7 t/ha to 1.9 t/ha, an increase of 3.2 percent per year. Guatemala is self-sufficient in maize production in most years, although imports in years of unfavorable weather can approach 10 percent of total consumption.

Despite the existence of much small-scale, subsistence-oriented maize production, 45 percent of the national maize area is found on farms larger than 7 hectares. In the major maize production region, along the Pacific Coast, farm sizes range between 7 and 45 hectares. About 75 percent of all maize is produced as a monocrop. The balance is produced in association with beans and other grain legumes or sorghum.

Direct human consumption accounts for 60 percent of total national maize production. The average Guatemalan consumes 350 grams of maize per day. Virtually every Guatemala farmer grows at least enough maize to satisfy family requirements.

Principal Maize Ecologies

Guatemala is located entirely within tropical latitudes. However, the elevation of the land area ranges from sea level to about 4,000 meters in the volcanic highland axis running west to east throughout the country. Rainfall ranges from 2,000 millimeters per year in certain lowland areas to less than 500 millimeters per year in some semi-arid valleys. Distinct wet and dry seasons exist in major crop production areas.

Maize is grown under three major environmental classifications in which eight maize production zones can be delineated (Fig. 10.5). The environments are
- Lowland tropical: less than 1,300 meters above sea level, where hybrids and varieties are planted in a 120-day growing season.
- Transitional: 1,300 to 2,000 meters above sea level
- Highland: above 2,000 meters, where the growing season is longer than 180 days.

The lowland tropical environments account for about 85 percent of the total maize area (Table 10.9).

The Maize Research System

Organized maize research began in the 1940s, with important support provided by the Rockefeller Foundation. During 1950s, Guatemala researchers participated in the collection of indigenous maize landraces,

FIGURE 10.5 Maize production zones in Guatemala.

under the auspices of the U.S. National Academy of Sciences-National Research Committee on Preservation of Indigenous Strains of Maize Project. In 1954, the Rockefeller Foundation initiated the Central American Corn Improvement Program, which provided germplasm, training, and technical assistance to maize researchers in the Ministry of Agriculture's research department. In addition, Iowa State University established a tropical maize breeding station at Tiquisate on the Pacific coast in 1955. Out of this breeding work came the lowland tropical composite variety Tiquisate Golden Yellow, which later was widely grown in Thailand under the name Guatemala and in Indonesia and the Philippines under the name Metro. (Subsequently, Tiquisate Golden Yellow became a key component in the development of Thai Composite #1, forerunner of Suwan-1.)

During the 1950s, Guatemalan maize breeders released a number of improved open-pollinated varieties developed through mass selection. In 1958, two improved highland varieties, Xela and San Marceño, were released. During the 1960s, seven lowland tropical varieties and one highland variety were also developed, based mostly upon germplasm supplied by Rockefeller Foundation maize improvement programs operating in Mexico, Colombia, and Central America.

TABLE 10.9 Characteristics of maize production zones in Guatemala.

	Production zones							
	1	2	3	4	5	6	7	8
Area, 000 ha	20	30	20	20	10	150	25	500
Environment	highland tropical	highland tropical	transition zone	transition zone	transition zone	lowland tropical	lowland tropical	lowland tropical
Grain type	yellow dent	yellow dent	white dent	yellow dent	yellow dent	white dent	yellow dent	white dent
Growing season	main	main	main	main	main	main	main	main
Maturity	extra late	late	late	late	extra late	int.	int.	int.
Moisture stress	rare	rare	sometimes	sometimes	rare	often	rare	rare
Soil type	acid	acid	normal	normal	normal	normal	normal	normal
Biotic stresses[a]								
Northern leaf blight	●●	●	●●	●●	●●	○	○	○
Ear rot	●	●	●●	●●	●●	●●	●●	●●
Southern rust	○	○	○	○	○	○	●	●
Common rust	○	○	●	●	●	○	○	○
Tarspot	○	○	●	●	●	○	○	○
Puccinia maydis	○	○	○	○	○	○	●	●
Southern leaf blight	○	○	○	○	○	●	●	●
Curvalaria spp.	○	○	○	○	○	●	●	●
Grain weevila	●	●	●	●	●	●	●	●
Earworm	○	○	●	●	●	○	●	●
Spodoptera spp.	○	○	●●	●●	●●	●●	●●	●●

aStress ratings: ○ Biotic stress not present in zone. ● Present but not of economic importance. ●● Some economic losses. ●●● Significant economic losses. ●●●● Severe economic losses. ●●●●● Maize cannot be grown unless a resistant variety is grown or chemical control is applied.

Source: CIMMYT.

In 1970, the Guatemala government reorganized its agricultural institutions, creating new centralized and decentralized institutions, including the Institute of Agricultural Sciences and Technology (ICTA)—all under the overall responsibility of the minister of agriculture. The General Directorate of Agricultural Services was primarily responsible for agricultural extension, the promotion of basic food production, irrigation development, and seed certification.

ICTA's Maize Breeding Research. The key features of ICTA's maize research program are integration of disciplines and a strong commitment to on-farm research. Research on experiment stations—largely plant breeding—is carried out by scientists assigned to the national maize program. These maize researchers, in turn, work closely with the regional production research teams charged with planning and conducting on-farm research activities.

ICTA has about 10 scientists involved in maize breeding at six experiment stations. The ICTA maize improvement program has made rapid progress in the developing high-yielding open-pollinated varieties and hybrids. CIMMYT has played an important complementary role by assigning maize breeders and agronomists to Guatemala since 1973.

ICTA maize breeders have made extensive use of germplasm supplied by CIMMYT (Table 10.10). During 1965-91, ICTA released 39 improved varieties and hybrids for lowland areas and 6 improved varieties for highland areas. Of the 45 varietal releases, 35 contained CIMMYT germplasm in their parentage.

In the ICTA system, researchers begin their investigations in each maize-growing area by launching a series of agronomic and socioeconomic surveys. Conducted by ICTA regional production research team members and headquarters social scientists, these surveys assess the circumstances and practices of farmers in target areas. Survey results allow researchers to screen and rank research priorities and to

TABLE 10.10 Tropical lowland maize varieties and hybrids containing CIMMYT germplasm in commercial use in Guatemala, 1990.

Variety name	Grain type	Grain color	CIMMYT source material
ICTA B-1	OPV	white	Tuxpeño-1 (Pop. 21)
ICTA B-5	OPV	white	Blanco Cristalino-2 (Pop. 30)
La Maquina-22	OPV	white	Mezcla Tropical Blanca (Pop. 22)
La Maquina-43	OPV	white	La Posta (Pop. 43)
HB-83	DC	white	Family of Pop. 22; inbred lines of Pops. 29 and 43
ICTA A-4	OPV	yellow	Amarillo Cristallino-2 (Pop. 31)
ICTA A-6	OPV	yellow	Amarillo Dentado (Pop. 28)
HA-46	DC	yellow	Family of Pop. 28 and inbred lines of Pops. 27
Nutricta	QPM/OPV	white	Quality Protein Maize Tuxpeño

OPV = open-pollinated variety. DC = double-cross hybrid. QPM = quality protein maize.

identify groups of farmers for whom a given technology—variety and crop management practices—should produce roughly similar results.

The maize research program has focused on the three major agroclimatic regions in which maize is produced—highland, transition, lowland. Research priorities differ from zone to zone. In most of the lowland agroclimatic region, agriculture is commercially oriented. Farmers generally are accustomed to buying agricultural inputs and to marketing their crops. To serve farmers in the lowland tropical production zones, ICTA's maize improvement work followed two closely related research lines. One strategy was to develop high-yielding varieties, whose seed could be saved by the farmer from year to year with little loss of yield vigor. To serve the commercially oriented farmers—many of whom were already using hybrids imported from neighboring El Salvador—ICTA maize breeders set out to develop Guatemalan hybrids. Here, the initial breeding priority was to develop nonconventional hybrids, such as inter-varietal and "family" hybrids.

Most of the maize in the transitional region, which ranges from 1,300 to 2,000 meters in elevation, is found in the eastern portions of the country near Honduras and El Salvador, and in the transitional areas between the central highlands and the northern lowland areas. The transitional zone can be further subdivided according to moisture availability into a part that suffers frequent drought stress and a part that does not.

For the highland region, ICTA has emphasized the development of open-pollinated varieties whose seed can be saved by farmers from year to year (Table 10.11). ICTA scientists have focused their efforts on the western highlands, where most cropland is above 2,000 meters elevation, and the central highlands, where most of the farmland lies between 1,500 and 2,000 meters elevation. Because of the substantial agro-climatic variation within each sub-region, ICTA scientists have moved preliminary germplasm development activities from experiment stations to farmers' fields.

ICTA's On-farm Research. ICTA has decentralized most of its crop management research activities to regional program centers. Within the ICTA research system, interdisciplinary production teams conduct farmer surveys, on-farm experiments, technology verification, and some training-related activities associated with transferring recommended technologies. Administratively, production research team members report to regional directors, although their maize-related activities are planned and executed in collaboration with the national maize research program staff.

Private Maize Research. Private research is on the rise in Guatemala and has been encouraged by policies of ICTA and the government that promote the development of a private seed industry. ICTA makes all of

TABLE 10.11 ICTA's improved highland
yellow grain open-pollinated varieties, 1990.

Variety	Suitable elevation(m)
Toto-Amarillo (ICTA 606)	2,300-2,700
San Marceño	2,000-2,500
Chivarreto (ICTA 612)	2,600-3,000
Don Marshall	1,500-2,000
V-301	1,500-2,000
V-302	1,500-1,700
V-304	1,500-2,000
Chanin	1,200-1,500

Source: CIMMYT and ICTA.

its maize germplasm available—at a fee—to private firms. This includes foundation seed of registered ICTA varieties and hybrids as well as the full range of inbred lines developed by ICTA plant breeders. At present two private companies are involved in maize research. In 1990, these companies' research budgets were equivalent to approximately 50 percent of ICTA's maize research budget.

Maize Seed Industry

Guatemala's national seed industry policy is built upon a close public-private partnership. The complementarity between local seed companies and ICTA is a key element in the rapid growth of the Guatemalan maize seed industry. In this partnership, the basic (foundation) seed of ICTA's maize varieties and hybrids is made available to private seed growers who produce the seed under the supervision of ICTA and Ministry of Agriculture seed specialists.

Commercial Seed Production. About 97 percent of the certified maize seed marketed in Guatemala is produced by private growers registered with the Department of Seed Certification. The ICTA seed unit produces the balance. About 40 percent of the total volume of improved maize seed is based upon ICTA genotypes and the remainder is proprietary hybrids registered by private companies.

Artisan Seed Production. While certified seed sales have achieved considerable market penetration among lowland tropical maize producers, little commercial seed is sold in highland areas, where most maize farmers have small landholdings and are subsistence producers. To deliver improved seed to them, ICTA and the General Directorate of Agricultural Services in 1987 initiated an artisanal seed production program. This project relies on extension officers and leaders from local farmers' associations to carry out small-scale seed production of improved varieties in areas not served by commercial seed outlets. ICTA seed unit staff train local extension officers and artisan seed producers in seed production, provide technical backstopping, and supply small quantities of foundation seed for seed multiplication.

Maize Production Constraints

ICTA researchers have delineated three levels of technology typical of maize production systems in Guatemala (Table 10.12). At the lowest level of technology, farmers plant local varieties at low population densities with little or no fertilizer (chemical or organic) and no chemical control of pests, diseases, and weeds. Average yields for this low-input technology are under 2 t/ha. At the intermediate technology level, farmers plant improved science-based varieties (or improved farmer-selected varieties) at optimum population densities with sub-optimal amounts of fertilizer and apply chemicals to control insects not diseases or weeds. At the high technology level, farmers plant improved hybrids or open-pollinated varieties at optimum population densities, apply recommended fertilizer doses, and use chemical control of insects, diseases, and weeds. Considering benefits, costs, and risk, the intermediate technology level is the most attractive for farmers in the lowland zones and the high technology level is the most attractive for farmers in the mid-altitude moist zones.

Prospects for the Maize Economy

Future maize demand in Guatemala is expected to increase at about the same rate as population, 2.8 percent per year. Although the demand for maize as a human food is expected to grow at less than the rate of population growth because per capita maize consumption tends to decline as personal incomes rise, the growth of demand for maize as a livestock feed is expected to outpace population growth. It is unlikely that the total maize area will increase more than 1 percent per year. Therefore, for Guatemala to maintain self-sufficiency in maize production, average national yields must reach 2.5 t/ha by the end of the century, a feasible target with available technology. The major yield gaps

TABLE 10.12 On-farm crop yields (t/ha) in Guatemala for different environments, cropping systems, and production technology.

Environment	Cropping pattern	Technology level		
		Low	Intermediate	High
Lowland	Monocrop	1.6	3.0	3.3
Mid-altitude (moist)	Monocrop	1.7	2.3	3.9
Highland	Multiple: maize/	1.1	2.1	3.9
	beans	0.2	0.4	0.5

Source: Schmoock and Manlio Castillo 1989.

that must be closed are found within the highland maize-producing zones, where low-level technologies are generally employed.

Brazil

Brazil, a country of continental dimension, is Latin America's most industrialized nation. It has a diversified economy, with well-developed industrial, trade, and agricultural sectors. The 1980s, however, were been difficult for Brazil. Saddled with a huge external debt (US$111 billion in 1990) and rampant inflation, the country has been struggling to get its economy in order.

Brazil's Agriculture

Brazil's vast land area encompasses many climates. Consequently, its major crops range from mainstays such as maize, sugarcane, and coffee to relative newcomers like soybeans and fruits. Maize is the most important cereal, followed by rice and wheat; and soybeans are the principal grain legume. Coffee is the leading export crop and sugarcane is an important industrial crop because it is used to produce ethanol for automobile fuel.

There are sharp differences in the development of Brazil's regions. The northeast is populated by small-scale farmers who contend with scarcity of land, chronic drought, poor soils, lack of improved technology, shortages of capital, and inadequate investments in infrastructure. In contrast, farmers in the south and southeast regions of Brazil benefit from favorable climates, abundant land, access to capital and improved technologies, and relatively well-developed infrastructure.

During the 1950s and early 1960s, agriculture was not a high priority

Brazil's agriculture in brief (1991).	
Population	151.4 million
Population growth, 1991-2000	1.4%/yr
GDP	US$414,061 million
Per capita GDP	US$2,940
Agriculture vs GDP	10%
Adult literacy	81% total (women: 80%)
Labor force (1989)	23% in agriculture
Total land area	846 million ha
Arable & permanent cropland	79 million ha
Cereal crop area	23 million ha
Irrigation	2.6 million ha

Sources: World Bank 1993; FAO 1993.

Brazil's maize economy in brief (1990-92).

Maize area		12.6 million ha
Maize yield		2.0 t/ha
Maize production		25.2 million t
Per capita utilization		169 kg/yr
Net maize imports, 1989-90,		653,000 t/yr
Maize grain color		0% white
Maize used as	human food	13%
	livestock feed	76%
Area planted to	unimproved varieties	43%
	improved OPVs	13%
	hybrids	44%

Source: CIMMYT.

in Brazil's development strategies. Even so, agriculture grew vigorously. Between 1955 and 1965, production of domestic food crops rose by 5.7 percent a year, of which 4.4 percent was due to increasing land area. Production of export and industrial crops (coffee, sugarcane, cotton, sisal) grew at 9.4 percent per year. Two-fifths of this growth was due to area expansion and the rest to yield increases.

Between 1966 and 1977, export agriculture grew at a remarkable 23 percent per year. Most of the increase was due to yield gains. Production of soybeans, the great success story of this period, rose 38 percent a year. Much of the land in the south previously planted to coffee was shifted to soybeans. Meanwhile, domestic food crops—such as maize, beans, rice, and wheat— only grew 3.8 percent a year, mostly from area expansion. Brazilian policy makers relied on the nation's extensive land resources to meet domestic food needs. However, Brazil's high rate of population growth and the rapid spread of export-oriented agriculture were making it more difficult to bring new land under the plow. As the cost of clearing new land, building roads, and providing other infrastructure increased, the alternative of raising yields on existing cropland became more attractive. In response, the government greatly expanded its investments in agricultural research.

The National Maize Economy

Maize covers half of the total area planted to cereals. During 1990-92, 25 million tons of maize grain were produced annually on nearly 13 million hectares. National yields averaged 2 t/ha. Brazil's farmers grow mostly yellow dent-flint grain types. Feed use account for 75 percent of maize utilization, food use for 14 percent, and other uses (industrial products, seed, and wastage) for the remaining 11 percent. About two-thirds of the national maize area is planted to hybrids and improved open-pollinated varieties. Fifteen percent of maize production occurs on farms smaller than 10 hectares in size, 40 percent on farms of 10 to 50 hectares, and 45 percent on farms larger than 50 hectares.

Principal Maize Ecologies

Maize in Brazil is produced in both tropical and subtropical environments (Table 10.13). Only 5 percent of the national maize area is irrigated and frequent drought stress is a problem on half the total maize area. Soil toxicity problems also affect about half the total maize area. Maize yields in the central and southern regions of Brazil (Fig. 10.6), which account for 90 percent of national production, average about 2.3 t/ha, while yields in the north and northeast are around 1 t/ha.

Among the biotic stresses, important diseases include northern leaf blight, yellow leaf blight, southern rust, ear rots, and downy mildew (in southern regions). The major insect pests include fall armyworm (*Spodoptera* spp.) and various stored grain insects.

The Cerrado in central Brazil is one continuous block (175 million hectares) with a tropical savanna ecology. The climate is tropical humid with a long hot dry season. Most of the soils in the Cerrado have a low pH, high phosphorus fixation capacity and aluminum saturation, and low water-holding capacity, which causes maize to be prone to drought stress. In 1989, maize was planted on about 2.75 million hectares in the Cerrado, with a total production of 6.6 million tons (26% of national production) and an average yield of 2.4 t/ha. Until the 1960s, the Cerrado was considered to be worthless for agriculture, except for strips of alluvial soils along streams. The chief agricultural enterprise was extensive cattle production. For grazing, the natural savanna-brush flora has poor digestibility and nutritive quality, resulting in low carrying capacity.

The Maize Research System

Maize research in Brazil dates back to work at the Agronomic Institute of Campinas in the 1930s. Local Cateto varieties were selfed and crossed with local yellow dent varieties, resulting in the first commercial hybrids (inter-varietal). During the 1940s, U.S. dent varieties were introduced, which showed much greater heterosis with Cateto. Thereafter, the popularity of semi-dents grew in Brazil. Inbreds from the U.S. and Argentina were introduced, but most of this germplasm was not sufficiently adapted to be of much use. The Cateto orange flints, the Tuxpeño dents, and a local dent, Paulista, have dominated Brazilian germplasm.

TABLE 10.13 Characteristics of maize production zones in Brazil.

	Production zones				
	1	2	3	4	5
Area, 000 ha	500	1000	2000	4700	3800
Environment	lowland tropical	lowland tropical	lowland tropical	lowland tropical	subtropical
Grain type	yellow dent & flint	yellow dent & flint	yellow dent & flint	yellow dent & flint	yellow dent & flint
Growing season	main	main	main	main	main
Maturity	early	int.	early	int.	late
Moisture stress	rare	sometimes	usual	sometimes	often
Soil type	normal	acid	normal	normal-acid	normal
Biotic stresses[a]					
Northern leaf blight	●●	●●	●	●●/●●●	●●
Southern rust	●	●●●	●	●●	●●
Yellow leaf blight	●	●●	●	●●	●●
Ear rot	●	●●	●	●●/●●●	●●●
Downy mildew	○	○	○	●	●
Sugarcane borer	●	●	●●	●●●	●●
Lesser cornstalk borer	○	●●●	●●	●●●	●●
Spodoptera spp.	●●●	●●●	●●●	●●●	●●
Heliothis spp.	●	●	●	●	●
Leafhopper	○	○	●	○/●●	○
Rootworm	○	○	○	●●	●●
Grain weevils	●●●	●●●	●●/●●●	●●●	●●●
Grain moth	●●	●●	●●	●●/●●●	●●/●●●
Agrotis cutworm	○	●	●	●●	●●

[a]Stress ratings: ○ Biotic stress not present in region. ● Present but not of economic importance. ●● Some economic losses. ●●● Significant economic losses. ●●●● Severe economic losses. ●●●●● Maize cannot be grown unless a resistant variety is grown or chemical control is applied.

Source: CIMMYT.

FIGURE 10.6 Maize production zones in Brazil.

During the 1950s and 1960s, Brazilian maize scientists made extensive use of exotic germplasm from the Caribbean flint and dent complexes to form new germplasm complexes that substantially increased
the yield potential of improved varieties and hybrids. Seed of the related Caribbean dent populations—Azteca, Maya, Piramex, Centralmex,
IAC-1—have been widely distributed in large regions from northern Rio
Grande du Sul to the Amazon Basin.

In 1973, the government created the Brazilian Corporation of Agricultural Research (EMBRAPA). Maize research is carried out by the National Maize and Sorghum Research Center (CNPMS), headquartered in
Sete Lagoas, Minas Gerais. The National Maize Research Program com-
prises 133 individual research projects (in 1991), executed throughout
Brazil, in collaboration with 30 state institutions, in a cooperative agricultural research system.

In the past 25 years, EMBRAPA has provided a major impetus to research aimed at the Cerrado. EMBRAPA scientists initiated a well-

coordinated and systematic interdisciplinary research program, integrating past knowledge and generating new research information and products. Soil fertility and toxicity research and interdisciplinary agronomic research led to soil management strategies involving liming and other soil amendments.

CNPMS has been successful in developing high-yielding maize varieties and hybrids. Breeding objectives include the development of genotypes with improved harvest index, resistance to major diseases and insects, greater nutrient use efficiency, tolerance to drought and soil mineral stresses, and improved protein quality. Outstanding genotypes developed by CNPMS include BR 106, the most widely grown variety in Brazil, BR 451, a white-grain quality protein maize variety, and BR 201, a hybrid that has tolerance to soils that have high levels of aluminum. At least 16 of CNPMS maize breeding populations trace their origin to CIMMYT. Ten open-pollinated varieties and four hybrids, covering 5 percent of Brazil's maize area, are derived from CIMMYT germplasm. BR 201, the first aluminum-tolerant hybrid is based largely on CIMMYT's maize germplasm.

This new generation of crop varieties is now moving into farmers' fields. Among them are high-yielding aluminum-tolerant maize varieties and hybrids. In addition, improved crop management systems are available that incorporate crop rotations and minimum tillage that leaves crop residues on the surface to facilitate moisture penetration and reduce runoff and erosion.

Maize Seed Industry

Brazil's maize seed industry is largely private. In 1990, the industry processed 150,000 tons of certified maize seed, enough to cover 70 percent of the national maize area. Hybrids account for 93 percent of the certified seed sold and improved open-pollinated varieties for the rest.

An association of 28 small and medium-size private maize seed companies, called Unimilho, produces hybrids developed by CNPMS. These companies generate 20 percent of national certified maize seed sales. They pay CNPMS royalties to produce its seed, and these royalties almost completely finance CNPMS maize-research operations. Most of Unimilho seed is sold in the northern maize-growing regions with tropical environments.

Unimilho is an interesting example of public-private cooperation. This association has helped many small companies to enter the seed market, and in time, it hopes to be large enough to fund proprietary research. The presence of Unimilho in the marketplace, especially since BR 201 was introduced in 1989, has led to general decline in prices for double-cross hybrids. The average price of BR 201 is now taken as the benchmark by other seed companies for pricing their hybrids. The increased competition resulting from the presence of Unimilho is seen as healthy, although some large private companies have complained that Unimilho receives implicit subsidies (its exclusive access to EMBRAPA maize varieties and hybrids), which permit it to use less than full-cost pricing. They contend that public lines should be available to all.

Large private seed companies, with their own proprietary hybrids, operate in the southern maize-growing regions, which have a subtropical environment. The biggest company, Agroceres, accounts for about 50 percent of national hybrid seed sales. In addition, major transnational companies, such as Cargill, Pioneer, and DeKalb, also operate in this region. Collectively these companies account for 30 percent of national certified seed sales.

Excessive production of maize seed is a recurring problem. The Brazilian seed industry has a 20 percent surplus. While the national maize area has remained relatively stable, the commercial demand for seed has declined, due to the weak economic conditions in the country. Some farmers are re-planting advanced generations of hybrids, and the use of improved open-pollinated varieties like the EMBRAPA variety BR 106 has also increased.

Maize Production Constraints

A few farmers in the Cerrado are starting to put all the pieces of the new crop production technologies together properly. The challenge for Brazil is to help all farmers acquire those skills. However, despite the weak infrastructure and the problems soils, this region produces 25 percent of Brazil's maize, rice, and soybeans. It also harvests 20 percent of the coffee, supports 40 percent of the nation's cattle herd, and produces 12 percent of the milk.

Continued development of the Cerrado will require substantial investments in infrastructure. The profitability of maize or other agricultural production in the Cerrado is limited by high transport and freight costs. To haul grain 1,500 kilometers by truck currently costs about US$40 per ton. Future developments of railroad and river transportation systems, along with continuing improvements in road systems, will greatly enhance the economics of maize production of this region.

About 2 million hectares of maize land, mainly in the northeastern portion of the country, frequently suffer from drought. This area needs high-yielding early maturing materials that tolerate drought stress.

Maize borers (*Spodoptera* spp.) are the major insect problem, especially in the lowland tropical area but also in the subtropical environments. Because no significant genetic resistance has yet been developed, the most common control method involves the use of insecticides. EMBRAPA has also developed a biological insecticide, Baculovirus, which is quite effective and environmentally friendly.

Prospects for the Maize Economy

The Brazilian maize economy appears to have a bright future. Maize prices in Brazil reflect the world price, and thus respond to real market forces. In 1992, Brazil produced a record 33 million tons of maize, with an average yield of 2.4 t/ha. By contrast, 1991 production was 28 million tons, with a yield of 2.1 t/ha. Higher yields demonstrate the growing intensification of production, as farmers use superior varieties, increased quantities of fertilizer, and improved crop protection methods.

Future growth in maize demand is tied to demand for livestock and poultry products because 75 percent of Brazil's maize is used as a feed. If Brazil's economic growth is strong, domestic demand for poultry, pork, and beef also will be strong, and maize demand could increase by 3 percent per year during the remainder of this decade, or about 3 million tons per year.

At present, Brazil is more or less self-sufficient in maize production. To develop an export market for maize, Brazilian farmers must become lower-cost producers through the continued adoption of yield-increasing, cost-reducing technologies. The commercial production areas in the southern portion of the country are the likeliest sources of maize for export. These areas, however, would have to compete with low-cost producers in the United States and Argentina. Thus, until domestic yields are increased and marketing systems become more efficient, Brazil will not become an important exporter, despite large areas that potentially can be brought into production.

11

Prospects for the Third World Maize Economy

Demand Outlook

Maize, with its multitude of food, feed, and industrial uses, its high genetic yield potential, and its wide range of adaptation, will experience increasing demand in the decades ahead. Population growth and income changes are the key influences. During the 1990s, world maize demand is expected to grow at about 2.6 percent per year. Most of the increase will come from the Third World. To meet this demand, maize production must increase by approximately 123 million tons to 600 million tons by the end of the century (Table 11.1).

In developing countries, food use of maize will grow at perhaps 1.5 percent a year, slower than population, but feed use could grow at 3.4 percent a year. Feed demand will be especially responsive to rising incomes in many developing countries. With strong economic growth, the feed share of total maize utilization could expand from 50 percent in 1989-91 to 60 percent by 2000. Overall during the 1990s, maize demand in the developing countries will likely increase by 3 percent a year, growing from 200 million tons to roughly 280 million tons. Growth in maize utilization will continue to be strongest in the middle-income and newly industrialized nations where standards of living have been improving.

TABLE 11.1 Projected growth in world maize demand.

Country group	1989-91 production (million t)	2000 projected demand (million t)	Change 1990-2000 (%)
Developed market economies	239	272	14
Eastern Europe & former USSR	37	49	32
Developing economies	202	280	39
World	477	600	26

Source: Authors' estimates.

In the former Soviet Union and Eastern Europe, assuming a successful transition to market-oriented economies, it is likely that maize demand will increase by 2 percent per year during the remainder of the 1990s.

Demand for maize in developed countries is expected to be strong. In the United States, the total food and nonfood demand for maize will reach an estimated 60 million tons by 2000 (compared with 34 million tons in 1991) in response to the growing consumer uses as sweeteners, ethanol, and various new products such biodegradable plastics and low-calorie fat substitutes.

Supply Prospects

Although there is room for planting more maize in some parts of the world, most of the additional 123 million tons of maize needed to meet global demand must come from higher yields, especially in areas where maize yields are substantially below their potential. The completion of the Uruguay Round of the General Agreement on Tariffs and Trade (GATT) and the North American Free Trade Agreement (NAFTA) should lead to some changes in world maize production patterns. With less protectionism, maize production and trade will be determined by comparative advantage. Real prices are likely to rise 10 to 15 percent, especially if environmental taxes are imposed upon farmers in industrialized countries who are heavy users of fertilizers and crop protection chemicals. Under NAFTA, Mexican maize price subsidies must be removed within a 15-year period, although the government intends to eliminate them sooner. Mexican consumers prefer white grain, which sells at a 15 to 20 percent premium over yellow maize. With this price premium, it is likely that U.S. producers and advanced Mexican farmers will expand their production of white maize, using high-yielding hybrids and crop management practices. Small-scale Mexican farmers who have relatively low-yielding production systems also will continue to produce white maize, but mainly for home consumption.

It is unlikely that yield levels in the developed economies of North America and Europe will increase appreciably during the 1990s. Instead, farmers probably will seek to reduce production costs at present yield levels. Significant research advances made in tillage systems, soil fertility management, and crop protection suggest that sophisticated farmers in the industrialized countries will be able to maintain high yield levels while significantly lowering production costs and enhancing environmental quality.

Ten million hectares in North America and Western Europe have been retired but could be returned to production, adding 60 to 70 million tons of maize annually to world supply. However, this scenario is not likely to materialize because GATT mandates lower maize subsidies. In the countries of the former Soviet Union and Eastern Europe yield levels could be raised by 30 to 50 percent, and several million hectares could be brought into production. Thus, maize production could increase by 10 to 15 million tons in these countries in the 1990s.

In developing countries of Asia, maize area is unlikely to expand significantly because population densities are high and arable lands are already intensively used. Latin America has little additional land to bring into maize production except perhaps for the Cerrado in Brazil. In sub-Saharan Africa, a considerable amount of unused land exists. However, the costs for transportation links and human settlements necessary to bring new areas into cultivation will be high, and expanding cultivated land often will have negative consequences for forests.

Even if the Third World maize area expands moderately (by 10 million hectares), yield levels must increase 20 to 30 percent to meet projected demand for this decade. At present, the average maize yield in developing countries is 2.4 t/ha. Yields must reach 3.0 t/ha by 2000 to maintain current levels of self-sufficiency, which is well within the realm of technical feasibility. Achieving these yield targets will depend largely on extending the use of high-yielding varieties and hybrids and on improving crop management practices, especially soil fertility management, particularly in countries where average maize yields are below 1.5 t/ha. Fortunately, improved genotypes and crop management practices already exist to substantially increase maize yields in such environments.

The Research Pipeline

Maize research progress has been a journey of ever-deepening knowledge about the genome. Scientific understanding of the laws and modes of genetic inheritance led to a quantum jump in maize research, resulting in the development of higher yielding varieties. Continued study of maize at the cellular level has led to new knowledge at the chromosome level. Now, maize scientists have delved into the molecular level and are exploring the most elemental functions of the maize plant. Molecular biology will enable scientists to do a better job of developing improved genotypes through conventional breeding. It also will help scientists to characterize and use genetic diversity in more productive ways.

At the same time, these advances will allow scientists to develop germplasm to overcome various biotic and abiotic stress problems for which solutions are expensive and time-consuming through conventional breeding techniques. Genes for increased tolerance or resistance to various biotic and abiotic stresses are present in the crop and related species. As knowledge of gene actions increases—and as scientists become better able to introduce alien genes into maize—new genotypes will be developed with markedly increased yield dependability. Biotechnologists are working with alien genes to confer insect resistance, e.g., the Bt gene is being introgressed into various maize genotypes that already have some level of resistance. Biotechnology tools will accelerate the development of varieties that possess resistance to herbicides, insects, and diseases. Herbicide-resistant maizes will give the farmer greater flexibility in planting patterns because herbicide carryover damage from other crops will be less likely. In the future, durable resistance to insects and diseases is likely to be achieved through a combination of conventional breeding and genetic engineering. We can also expect that through the tools of genetic engineering, maize varieties will be developed that have a better balance of amino acids, i.e., higher levels of lysine, tryptophan, etc. Genetic engineering will accelerate the development of maizes for industrial uses, such as grains with altered starches and oils and varying degrees of endosperm hardnesses. Genetic engineering will also allow scientists to create a range of seed-coating treatments that act as a carrier for fungicide, insecticide, and mycorrhizae complexes that will make the micro-environment around the planted seed more favorable. The use of genetic carriers, such as endophytes, also affords a mechanism to introduce resistance to insects and other biotic stresses.

Despite the promising outlook for maize research and development in the next 10 to 20 years, the major sources of improved germplasm will continue to come from conventional plant breeding programs. Excellent maize germplasm is available for the lowland and intermediate elevations of the tropics and subtropics. Only for the highland areas has germplasm development lagged behind, and even there germplasm is rarely the major limiting production factor.

Intellectual Property Rights and Technology Access

A guiding principle at CIMMYT and IITA has been the unrestricted exchange of germplasm and information, thereby promoting the widest possible benefit for farmers and consumers. With the rapid development of private maize research and seed production throughout the world, the issue of legal protection for intellectual property rights, including plant breeder's rights and plant variety protection, has become a matter of concern to international maize research programs. CIMMYT and IITA (and some advanced national programs in the developing world) are also carrying out biotechnology research to enhance their maize improvement efforts and to help developing countries gain access to new technology and products. As part of this work, the international centers may have to enter into contractual agreements with private companies that are much more advanced in the field of biotechnology research and that can be valuable sources of information and new technology. These agreements will undoubtedly lead to some restrictions being placed on the transfer and use of proprietary maize technology. In addition CIMMYT and IITA may be compelled to protect their own innovations, simply to be able to guarantee that they remain freely accessible to themselves and their clients.

The matter of intellectual property rights, plant breeder's rights, and plant variety protection has additional significance for maize research programs in the developing world as government research budgets shrink. By invoking plant breeder's rights and plant variety protection public-sector institutions hope to derive income from the sale of their germplasm products to augment their research budgets. Moreover, some leaders of publicly funded maize research programs feel that private organizations, rather than getting public-sector research products free of charge, should share some of their profits from using them. Thus, the effect of intellectual property rights is likely to restrict the historically open institutional nature and culture of most public-sector institutions.

Proper handling of complex intellectual property rights issues is crucial for establishing productive complementary relationships between public and private research organizations. Publicly funded national programs that want to introduce such regulatory mechanisms needed to develop sound policies, rules and regulations for variety nomenclature, and procedures for registration and certification. And they should recognize that the costs of formulating, regulating, and defending intellectual property rights can be high and sometimes prohibitive. Some private maize research institutions are seriously considering whether to bother about intellectual property rights at all or simply focus on rapidly developing new products to replace old products on the market. Probably both strategies will be used, as appropriate.

Prime Factors of Maize Development

Substantial investments will be needed in infrastructure development, agricultural research and extension, input production and distribution systems, grain storage facilities, and marketing to capitalize on the large unexploited maize production potential of the Third World. For these investments to produce maximum returns, they must be complemented by economic policies that stimulate increased agricultural productivity in ways consistent with the wise use of natural resources.

Infrastructure Development

Characteristically, modern maize economies consume large amounts of inputs and produce large production surpluses that must be moved through organized markets. An efficient transportation system is vital to link farm input suppliers, maize growers, and consumers in integrated production and marketing systems. The improvements needed include all-weather farm-to-market roads for transporting inputs and produce and accessible warehouses to store grain. In sub-Saharan Africa, transportation costs, which are twice as high as in Asia and Latin America, are a critical constraint to agricultural development. Sub-Saharan Africa also has inadequate facilities for storing surplus grain. As a result, the price of maize fluctuates violently. It tends to collapse at harvest time and skyrocket in the months immediately before the next harvest. A network of storage facilities with a capacity of 15 to 20 percent of annual grain consumption would help stabilize prices as more maize enters commercial market channels. Greater storage capacity would also improve the food security of rural areas.

Research and Technology Generation

History has shown that the maize economy operates best as an integrated research and production system involving both private and public organizations. For maize research and technology generation, a mixed international-national research system involving public and private organizations is most effective. In such a system, public organizations should focus on maize research topics that have broad applicability. Research products that emerge from publicly funded research should be made available to the private sector, with any royalty fees applied universally and set at accessible levels. The research products can then be used by private maize seed companies for marketing directly to farmers or as intermediate products in proprietary research programs. However, we believe that seed production, input supply, and grain marketing should be undertaken largely by private organizations, with governments mainly concerned with ensuring that markets operate in efficient, responsible, honest, and safe ways.

There is considerable need for more crop management research, especially on soil fertility management, weed control, and integrated pest management. In particular, minimum-tillage land preparation and weed control technologies should be developed and introduced. These technologies can dramatically reduce labor requirements (per unit of cultivated land), improve moisture conservation (mulch effect), and diminish soil erosion. All of these improvements will benefit most small-scale, resource-poor farmers. In addition improved soil fertility practices, including use of green manure, are urgently needed in most soils. Unless such improved crop management technology is generated and made available to small-scale farmers, progress in modernizing maize production systems will be stymied in many countries.

Extension Education

Most developing countries have established agricultural extension services to engage in various agricultural and rural development activities. The extension services are charged with training farmers in the methods of new production technologies. Extension agents also help to coordinate on-farm field testing and technology evaluations. Their work involves selecting farmer cooperators, providing technical training in the new technology, managing the logistics of supplying the inputs needed for demonstrations, and conveying information to maize researchers about key production problems facing farmers as well as the performance of the technological packages being recommended.

The impact of extension services in the transfer of improved technology is a matter of debate. Some development specialists contend that if the technology is clearly superior and appropriate, it will rapidly diffuse, whether an organized extension effort exists or not. Often, farmers learn about a new variety or crop management system through informal information channels—such as a progressive farmer who is a family member, friend, or neighbor. Even so, a technically competent and responsive extension organization can surely improve the diffusion of skills needed in modern maize production practices and thus accelerate the adoption of improved technology. Many small-scale farmers who stick with traditional technology know about the potential benefits of improved technology. However, their traditional crop management skills do not prepare them for the management requirements of higher yielding technologies. In particular, farmers may not recognize the importance of timely fertilizer applications, the differing effect of the various nutrients, nor the need for early weeding when soil fertility is improved.

Maize extension activities in developing countries—especially involving small-scale farmers—will likely remain publicly funded activities for some time. Private companies that offer seed and crop protection chemicals will carry out some extension activities in conjunction with promoting their products. However, because small-scale farmers will be modest consumers of inputs, it will not pay private companies to offer them substantial services. (Private-sector extension traditionally has been tied to cash crops such as tobacco, cotton, and sugarcane where farmers must deliver their output to the company providing the technical assistance; thus, the cost of extension is paid through the price of inputs supplied and through the output prices paid to the farmer.)

Input Supply

In maize, two inputs—improved seed and fertilizer—are of overriding importance to the modernization process. In most developing countries, parastatal government organizations were set up to produce and distribute seed and to provide various services such as input supply and traction power to farmers. However, public-sector agricultural industries and services have rarely operated at a profit or functioned with sufficient efficiency to develop a strong demand for their products among farmers. In recent years, the private sector has become increasingly involved in the supply of improved maize seed, fertilizers, and other agricultural chemicals and equipment in most developing countries. In order for the private agri-business sector to develop and flourish, publicly funded organizations must avoid engaging in unfair competition. Governments should ensure that their investments and activities in maize research and development support and complement the work of private entrepreneurs engaged in the delivery of improved maize technology. Moreover in most developing countries, government mechanisms for maize variety evaluation and release are very inefficient. These regulatory systems must be be streamlined and made more transparent.

The use of improved maize seed—especially hybrids—and use of fertilizer are usually closely correlated. Without improvements in soil fertility, little improvement in yield is possible. Once soil fertility is raised, the use of management-responsive varieties becomes much more profitable. Farmers generally begin applying some fertilizer before beginning to purchase certified seed. In the initial stages, when most chemical fertilizers are imported, the problem facing agricultural planners is how to develop a fertilizer supply system that can acquire the right kinds and amounts of fertilizer products and deliver them to the farmer in a timely and efficient manner.

Human Capital and Managerial Skills

Many maize farmers in low-income countries cultivate small landholdings, are impoverished, nearly illiterate, and subject to a poorly functioning marketing system. Despite these drawbacks, small-scale maize farmers are shrewd operators and capable of utilizing more advanced production technologies. In the early stages of maize development, farmers can quickly grasp the new practices of the recommended technological packages. As maize production systems become more intensive, however, the successful use of higher yielding technologies requires more sophistication. A maize farmer's education obviously affects his or her access to written information, understanding of technical concepts, and ability to manage more costly production systems. Maize farmers who lack reading and mathematics skills, therefore, are increasingly at a disadvantage. Thus, the importance of government investments to improve general rural education systems cannot be stressed enough. The linkages between levels of education, adoption of improved technology, and reduction of poverty are clear-cut. Greater educational opportunities for women are especially important, not only because of their role in maize production but also because of the strong correlation between higher education levels and smaller families.

Producer Incentives

Although the adoption of modern maize technologies offers the potential for greater economic benefits, it also exposes the farmer to greater financial risks than lower yielding traditional systems of production. Economic incentives, therefore, are required to induce maize farmers to adopt improved technologies. In simplest terms, farmers must have good prospects to make a profit if they invest in more modern production systems. Governments must find ways to ensure that the cost:price relationships between inputs and outputs provides a reasonable rate of return to the investments being made in modern maize production inputs and agro-services (potential risks from crop failure also must be factored into these calculations). Input subsidies or grain price supports are possibilities. Also, improvements in agricultural marketing systems are needed to protect the farmer against price collapse after harvest and to keep the government from having to invest in large carryover stocks.

Agricultural Credit

A fundamental challenge to the intensification of maize production among resource-poor farmers in the Third World is how to help them finance the purchase of recommended inputs. It is unrealistic to think that farmers with less than 2 hectares can be served by commercial banking institutions, because the bank transaction costs are too great in relation to the size of an average loan, which is likely to be under $200. Further, many small farmers, especially in sub-Saharan Africa where tribal ownership often prevails, do not have legal title to their land nor any other suitable collateral. For very small-scale farmers, production credit must come from the informal credit market or from family savings. However, for other small-scale farmers—those with the potential to plant 5 to 10 hectares—who need to borrow perhaps $1,000 in order to adopt higher yielding maize technologies, institutional innovations will have to occur in traditional banking to lower transaction costs. These innovations include (1) a village-based approach to lending rather than a branch-based approach, (2) coupling technology transfer with provision of credit, (3) reliance on yield potential and risk of recommended maize technology rather than on collateral in determining whether to grant loan, and (4) insuring that loans are disbursed in a timely and cost-effective way.

Summing Up

The tremendous range of uses for the maize plant, its adaptation to many environments, and rapid research advances that have made maize the most productive cereal, all suggest a bright future for the maize crop in coming decades.

There is considerable debate today about whether the adoption of modern production inputs, such as fertilizers, crop protection chemicals, and hybrid seed, is practical for small-scale, resource-poor maize producers—especially in sub-Saharan Africa. Some contend that low-income countries cannot afford to use scarce foreign exchange to import the modern agricultural inputs and equipment. Others question the suitability of using fertilizers and crop protection chemicals on environmental safety grounds.

We believe that the adoption of improved maize technology by all maize farmers—large and small—is a necessary condition for profitable production. Nearly half of the Third World maize area has low fertility soils planted to unimproved genotypes. In most tropical and subtropical environments, farmers can easily double maize yields with the use of modern varieties, improved soil fertility, and effective control practices for weeds, diseases, and insect pests.

It would be unrealistic to advocate that small-scale, resource-poor maize producers aspire to the high-input, capital-intensive production systems found in the United States and Europe. But we are advocating that all Third World maize farmers use modern genotypes with the best-available genetic resistance to diseases, insects, and abiotic stresses of economic consequence, apply moderate amounts of fertilizer (50 to 100 kg/ha of nutrients), and practice effective weed control, often through the application of herbicides. Small-scale farmers, it is hoped, will use these modern inputs in combination with such crop management practices as green manures, minimum tillage systems, and integrated pest management, which can save on the expense of using purchased inputs and contribute to environmental protection. Our experience has demonstrated that Third World maize farmers—including small-scale, resource-poor producers—are capable of using modern technological components—especially improved seed, fertilizers, and land preparation and weed control systems using farm chemicals.

It is important that those working in maize research and development maintain balance when talking about sustainable maize technologies for the Third World. Without the adoption of modern inputs, the low-yielding maize production systems of millions of small-scale farmers will never be transformed, and the pervasive and debilitating poverty afflicting these producers will continue.

References

American Society of Agronomy. 1978. *Multiple cropping*. Special Publication 27. Madison, Wisconsin.

Bolaños, J., and G. O. Edmeades. 1993. Eight cycles of selection for drought tolerance in lowland tropical maize: Responses in grain yield, biomass, and radiation utilization. *Field Crops Research* 31:233-52.

Bosque-Perez, N. A., and J. H. Mareck. 1990. Screening and breeding for resistance to the maize stem borers, *Eldana saccharina* and *Sesamia calamistis*. *Plant Resistance to Insects Newsletter* 16:119-20.

Brewbaker, J. L., M. L. Logrono, and S. K. Kim. 1989. *The MIR (maize inbred resistance) trials: Performance of tropical-adapted maize inbreds*. Research Series # 62. Honolulu: College of Tropical Agriculture and Human Resources, University of Hawaii.

Byerlee, D. 1987. *Maintaining momentum in post-green revolution agriculture: A micro-level perspective from Asia*. MSU International Development Paper No. 10. East Lansing Michigan: Michigan State University.

CIMMYT (International Maize and Wheat Improvement Center). 1987a. *1986 CIMMYT research highlights*. Mexico City.

CIMMYT. 1987b. *1986 World maize facts and trends*. Mexico City.

CIMMYT. 1988. *Recent advances in the conservation and utilization of genetic resources*. Mexico City.

CIMMYT. 1992. *1991/92 World maize facts and trends*. Mexico City.

CIMMYT. 1994. *1993/94 World maize facts and trends*. Mexico City.

CIMMYT Maize Program. 1988. *Maize production regions in developing countries*. Photocopy. Mexico City.

Darrah, L. L., and M. S. Zuber. 1986. 1985 United States farm maize germplasm base and commercial breeding strategies. *Crop Science* 26:1109-13.

De Leon, C., G. Granados, and R. N. Wedderburn, eds. 1988. *Proceedings of the Third Regional Maize Workshop, June 8-15, 1988, Kunming and Nanning, Peoples Republic of China*. Bangkok: CIMMYT.

De Leon, C., and S. Pandey. 1989. Improvement of resistance to ear and stalk rots and agronomic traits in tropical maize gene pools. *Crop Science* 29:12-7.

Douglas, Johnson. 1980. *Successful seed programs: A planning and management guide*. Boulder, Colorado: Westview Press.

Edmeades, G. O., J. Bolaños, and H. R. Lafitte. 1990. Selecting for drought resistance in maize adapted to the lowland tropics. In *Proceedings of Fourth Asian*

Regional Maize Workshop, Islamabad, Pakistan, September 22-28, 1990, ed. C. De Leon, G. Granados, and M. D. Read. Islamabad: CIMMYT.

FAO. 1993. *Agrostat-PC* [computer disks]. Rome.

Fischer, K. S., G. O. Edmeades, and E. C. Johnson. 1987. Recurrent selection for reduced tassel branch number and reduced leaf area density about the ear in tropical maize populations. *Crop Science* 27:1150-6.

Galiant, Walter C. 1988. The origin of corn. In *Corn and corn improvement*, ed. G. F. Sprague and J. W. Dudley, 3–30. 3rd ed. Madison, Wisconsin: American Society of Agronomy.

Geraldi, L. O., F. J. B. Biranda, and R. Vencovsky. 1985. Estimates of genetic parameters for tassel characters in maize (*Zea mays* L.) and breeding perspectives. *Maydica* 30:1.

Goodman, M. M. 1985. Exotic maize germplasm: Status, prospects, and remedies. *Iowa State Journal of Research* 59:497-527.

Gracen, V. E. 1986. Sources of temperate maize germplasm and usefulness in tropical and subtropical environments. *Advances in agronomy*, vol. 39. New York: Academic Press.

Harrington, L., S. Whangthongtham, P. Witowat, R. Meesawar, and S. Suriyo. 1991. *Beyond on-farm trials: The role of policy in explaining non-adoption of fertilizer on maize in Thailand*. Presented at the 11th AFSRE Symposium, Michigan State University, East Lansing, Michigan.

Johnson, E. C., K. S. Fischer, G. O. Edmeades, and A. F. E. Palmer. 1986. Recurrent selection for reduced plant height in lowland tropical maize. *Crop Science* 26: 253-60.

Khehra, A. S., B. S. Dhillon et al. 1988. Winter Maize Breeding and Production in the Indian Punjab. In *Proceedings of the Third Asian Regional Maize Workshop, Kunming and Nanning, China, June 8-15, 1988*, ed. Carlos De Leon, Gonzalo Granados, and Richard N. Wedderburn. Bangkok: CIMMYT.

Kim, S. K. 1990. Breeding of temperate maize germplasm for tropical adaptation. In *Proceedings of Fourth Asian Regional Maize Workshop, Islamabad, Pakistan, September 22-28, 1990*, ed. C. De Leon, G. Granados. and M. D. Read. Mexico City: CIMMYT.

Kim, S. K. and A. R. Hallauer. 1989. Agronomic traits of tropical and subtropical inbreds in Iowa. *Plant Varieties and Seeds* (No. 2): 85-91.

Kim, S. K., Y. Efron, F. Khadr, J. Fajemisin, and M. H. Lee. 1989. Registration of 16 maize-streak virus resistant tropical maize parental inbred lines. *Crop Science* 27:824-5.

Li Jingxoing. 1988. Hybrid maize breeding in China. In *Proceedings of the Third Asian Regional Maize Workshop. Kunming and Nanning, China, June 8-15, 1988*, ed. Carlos De Leon, Gonzalo Granados, and Richard N. Wedderburn. Bangkok: CIMMYT.

Linneman H., J. De Hoogh, M. A. Keyser, and H. D. J. Van Heemst. 1979. MOIRA: Model of international relations in agriculture. In *Potential world food production*. Amsterdam: North Holland Publishing.

Mareck, J. H., N. A. Bosque-Perez, and M. S. Alam. 1989. Screening and breeding for resistance to African maize borers. *Plant Resistance Newsletter* 15:58-9.

McMullen, N. 1987. *Seeds and world agricultural progress.* Report 227. Washington, D.C.: National Planning Association.

Mihm, J. A. 1989. Mass rearing stem borers, fall armyworms, and corn earworms at CIMMYT. In *Toward insect resistant maize for the Third World.* Mexico City: CIMMYT.

Mihm, J. A. 1990. Progress in breeding for host plant resistance to insects. Mexico City: CIMMYT.

National Corn Growers Association and National Corn Development Foundation. 1992. *The world of corn: A comprehensive look.* St. Louis.

Paterniani, E. 1990. Maize breeding in the tropics. *Critical reviews in plant sciences* 9(2): 125-54.

Paterniani, E, and M. M. Goodman. 1977. *Races of maize in Brazil and adjacent areas.* Mexico City: CIMMYT.

Plucknett, D. L. 1992. Modern crop production technology in Africa: The conditions for sustainability. In *Africa's agricultural development: Can it be sustained?* ed. N. C. Russell and C. R. Dowswell. Mexico City: Sasakawa Africa Association/Global 2000/Centre for Applied Studies in International Negotiations.

Plucknett, D. L., N. J. N. Smith, J. T. Williams, and N. M. Anishetty. 1987. *Gene banks and the world's food.* Princeton, New Jersey: Princeton University Press.

Pray, C. E., and R. Echeverria. 1991. Private sector agricultural research in less developed countries. In *Agricultural research policy: International quantitative perspectives,* ed. P. G. Pardey, J. Rosebloom, and J. R. Anderson. Cambridge, U.K.: Cambridge University Press.

Renfro, B. L. 1985. Breeding for disease resistance in tropical maize. In *Breeding strategies for maize production improvement in the tropics,* ed. A. Brandolini and F. Salmini. Publication no. 100. Firenze, Italy: FAO

Renfro, B. L. Unpublished. Genetic resistance to diseases in maize. Mexico City: CIMMYT.

Schmoock, W., and Luis Manlio Castillo. 1989. *Perfil del maiz, Guatemala: Sus implicaciones en el establecimiento de prioridades de investigacion.* Guatemala City: CIMMYT.

Shaw, Robert H. 1988. Climate requirement. In *Corn and corn improvement,* ed. G. F. Sprague and J. W. Dudley, 609–38. 3rd ed. Madison, Wisconsin: American Society of Agronomy.

Trifunovic, V. 1978. Maize production and maize breeding in Europe. In *Maize breeding and genetics,* ed. D. B. Walden, 41-58. New York: John Wiley & Sons.

Watson, Stanley A. 1988. Corn marketing, processing, and utilization. In *Corn and corn improvement,* ed. G. F. Sprague and J. W. Dudley, 885-940. 3rd ed. Madison, Wisconsin: American Society of Agronomy.

Wellhausen, E. J. 1978. Recent developments in maize breeding in the tropics. In *Maize breeding and genetics*, ed. D. B. Walden, 56-84. New York: John Wiley & Sons.

World Bank. 1993. *World Development Report 1993*. New York: Oxford University Press.

About the Book and the Authors

Maize is the world's most widely grown cereal and a dietary staple throughout the Third World, but its full potential has only begun to be tapped. This book thoroughly examines the biological and economic issues relevant to improving the productivity of maize in developing countries. The authors explore a wide range of practical problems, from maximizing maize's reproductive behavior and genetic resistance to insects and diseases comma deleted to identifying research priorities, promoting private-sector services, and utilizing technology effectively. They also examine the place of maize in the world economy, the roles of public and private research organizations, and problems of policy.

Christopher R. Dowswell is director for program coordination of the Sasakawa Africa Association. **R. L. Paliwal** and **Ronald P. Cantrell** are former directors of the maize program of the CIMMYT (International Maize and Wheat Improvement Center) in Mexico City.

Index

Printed and bound by CPI Group (UK) Ltd, Croydon, CR0 4YY

23/10/2024

01778232-0002